STATISTICAL MECHANICS
An Introduction

STATISTICAL MECHANICS
An Introduction

D. H .TREVENA
Department of Physics
The University College of Wales
Aberystwyth

Horwood Publishing
Chichester

First published in 2001 by
HORWOOD PUBLISHING LIMITED
International Publishers
Coll House, Westergate, Chichester, West Sussex PO20 6QL
England

Reprinted 2003

Printed and bound by Antony Rowe Ltd, Eastbourne

British Library Cataloguing in Publication Data
A catalogue record of this book is available from the British Library

ISBN: 1-898563-89-6

Table of contents

Foreword

Equilibrium statistical mechanics has the unjust reputation of being abstruse and 'messy'. Yet as far back as 1936 my teacher, R. H. Fowler, was able to write 'No difficulties of principle remain' in the second edition of his great book *Statistical mechanics*. Indeed, the advent of quantum theory, with its insistence that the energy levels of a closed system are distributed discretely rather than in a continuous spectrum, showed us how to simplify very considerably the basic mathematics of equilibrium statistical mechanics in ways that are explained in the book.

There is a real need for a book to bring this good news to students, also to show them that models exist whose consequences can be worked out exactly, but which are not too far removed from reality. Examples are the perfect gas and electrons in a metal. I still remember my satisfaction when I finally realized that the vast differences in the conductivities of metals and insulators could be accounted for by comparatively simple considerations. Now one can add that the basic physics of 'semiconductor devices' are well within the reach of an honours student as are the basic physics of polymers.

I agree entirely with Dr Trevena that there is a real need for a book to be used mainly by honours students and that many of the advanced books on the subject tend to frighten students away from a fascinating and very useful branch of physics.

I wish the book every success.

H N V Temperley Sc.D.
Emeritus Professor of Applied Mathematics, Swansea
Former Fellow of King's College, Cambridge
Royal Society Rumford Medallist and former Smithson Research Fellow

Preface

There have been many books written on statistical mechanics, mostly of an advanced nature and, in my experience, too difficult for the typical present-day undergraduate. Whilst preparing and delivering lectures on the subject for over thirty years to such students it became increasingly clear to me that there was a need for a concise introductory book on the topic which could be easily read and understood by second and third year undergraduates in physics, applied mathematics, physical chemistry, chemical engineering, metallurgy, materials science and polymer science. It was with such readers in mind that I have written this book but I hope also that it will be useful to research workers in higher education, government establishments and industry.

The book starts with a brief historical introduction after which the Maxwell–Boltzmann, Fermi–Dirac and Bose–Einstein statistics are treated, as is their application to gases and solids, including metals and semiconductors. Two novel features of the book are a chapter on liquids, an area largely ignored previously except in the advanced treatises, and also a chapter devoted entirely to worked examples. The mathematics throughout has been kept as simple as possible.

I wish to express my thanks to various friends who have helped me in the preparation of this book. My greatest debt is to Professor H. N. V. Temperley who has patiently read through the whole manuscript and suggested several improvements. My association with him goes back to 1950 when I did research under his direction at Cambridge and I owe much to the encouragement and friendship that he has given me over the years. I also wish to thank various colleagues in the Physics Department at Aberystwyth: Professor D. H. Edwards, Dr D. F. Falla, Dr Tudor Jenkins, Dr Iestyn Morris and Dr Eleri Pryse read parts of the manuscript and made useful suggestions while Professor Lance Thomas and Dr G. O. Thomas gave me a *pied-à-terre* in the department to do some writing after I had taken early retirement. I also thank Noreen Davies for her efficient work in typing the whole manuscript. I am also most grateful to Mr Ellis Horwood and his colleagues for their help and cooperation at all stages during the preparation of this book for publication.

D. H. T.

1

Historical introduction

Broadly speaking, this book deals with the application of equilibrium statistical mechanics to gases, solids and simple liquids. The history of statistical mechanics is intertwined with that of thermodynamics and kinetic theory. All three, which developed concurrently, amount to entirely different ways of dealing with properties of matter. We are considering the basic problem of accounting for the properties of an assembly of molecules and we have three distinct but related approaches. Thermodynamics deduces relations *independent* of the properties of the molecules and mechanisms of their interactions. Kinetic theory attempts to trace the average history of a typical molecule, while statistical mechanics builds on the general result that the equilibrium properties of an assembly are, in principle, known if the energies of all possible microstates are known.

The basic ideas were developed mainly during the second half of the nineteenth century and very early twentieth century. However, some of the preliminary groundwork was laid much earlier and was concerned with what was essentially the kinetic theory of perfect gases. We shall start our story with the work of Daniel Bernoulli in his *Hydrodynamica* in 1738. Being aware of the atomic nature of matter he assumed that the pressure of a gas was due to the impacts of the gas particles on the enclosure walls and he succeeded in deducing Boyle's law. After this, little seems to have happened for about a century until the work of Waterston, Clausius and Clerk Maxwell, whose first paper on molecular velocities was read in Aberdeen in 1859; in it the famous Maxwell distribution of molecular velocities in a gas in equilibrium was presented.

Soon afterwards Boltzmann extended Maxwell's theory and showed that with

Fig. 1.1. A photograph of Ludwig Boltzmann's tombstone showing his famous equation. (By courtesy of Professor Dieter Flamm of the University of Vienna.)

every degree of freedom of a gas molecule there is associated the same mean energy. Boltzmann's great contribution was to give the relation between entropy and probability and the famous equation $S = k \ln W$ appears on his tombstone in Vienna (see Fig. 1.1). This work laid the foundation of what we now call classical or Maxwell–Boltzmann statistics. It should be noted that Willard Gibbs had also suggested a relationship between entropy and probability, so that we may really regard both Boltzmann and Gibbs as the two founders of statistical mechanics.

In 1873 van der Waals proposed his famous equation of state, which allows qualitatively both for the finite size of the molecules and for their mutual attraction; in other words the van der Waals equation describes an imperfect gas. Since the attraction is expected to become more important as the density increases, the theory predicts that at any temperature below a certain critical temperature T_c compression of an imperfect gas to a certain density is followed by a 'landslide'; the assembly collapses to a larger density at which the molecules are nearly in contact. We identify this with condensation to a liquid. These simple corrections to the 'perfect gas' law are capable of accounting qualitatively, not only for the observed departure of real gases from perfection, but also for the observed phenomena of condensation, the critical temperature and even some of the properties of the liquid phase as well.

At the turn of this century, with the advent of the quantum hypothesis, Planck used statistical methods to treat black-body radiation as a photon gas. This was followed in 1907 by Einstein's work on the heat capacity of solids; this was later modified by Debye and others. A great advance was made in 1924 when the combination of quantum and statistical ideas led to the treatment of a photon gas by the Bose–Einstein statistics. When, however, it came to treating a gas of *electrons* the restrictions imposed by the Pauli exclusion principle led to the appearance of the Fermi–Dirac statistics in 1926 and Sommerfeld's electron theory of metals in 1928. It was shown that the Bose–Einstein and Fermi–Dirac statistics were related, respectively, to the symmetry and antisymmetry of the wave function for the assembly of the corresponding particles. It was also shown that the Maxwell–Boltzmann distribution was the limiting case, at high temperatures, of the Bose–Einstein and Fermi–Dirac distributions. Then, in the 1940s, the unusual properties of liquid ^{4}He and liquid ^{3}He were explained by treating them as Bose–Einstein and Fermi–Dirac fluids respectively.

Again, in the early 1910s, von Laue and his colleagues and, later, W. L. Bragg, studied the structure of crystalline solids using X-ray diffraction. The regular pattern produced by the diffracted rays showed that the molecules were situated in a regular lattice arrangement and we therefore say that a crystalline solid exhibits long-range order. The next step was to examine liquids by X-rays; this was done by Debye and Scherrer in 1916 and in this case the diffraction pattern showed that a liquid does not possess long-range order but only local or short-range order. To describe this short-range order the concept of the radial distribution function was introduced in the 1920s. This quantity is the variation of local density that would be measured by an observer at the centre of a typical molecule and it has been shown that X-ray and neutron scattering are closely related to this quantity, which is the analogue of the crystalline structure of a solid. As a result a liquid can be regarded as a disordered solid where only short-range order exists. We sometimes say that the molecules of a liquid form a quasi-crystalline arrangement and these ideas form the basis of one approach to the study of the structure of a liquid.

We now return to consider 'real' chemical gases (as distinct from a gas of electrons or photons). After the early work of the second half of the nineteenth century, described earlier, the theory of gases was further developed by Jeans and then by Lennard-Jones during the early decades of this century. This was followed by a study of dense gases and the elegant theory of the Mayers. In this way another approach to the study of liquids was conceived, namely, that of treating a liquid as a dense gas. Significant contributions to the theory of liquids were made by Born and Green, Kirkwood, Percus, Yevick, Rushbrooke, Temperley and others.

In addition, statistical mechanics has been applied to non-equilibrium phenomena such as transport properties. Another area has been that of phase transitions and the critical region, topics so comprehensively treated in Temperley's book *Changes of state*. These matters are, however, outside the scope of the present book.

2

Some basic ideas

2.1 INTRODUCTION

The methods of statistical mechanics enable us to predict the properties of matter in bulk in terms of the behaviour of the basic particles of which it is composed. In other words they are a 'bridge' which takes us from a microscopic description of matter to the equilibrium macroscopic thermodynamic description. This latter description of a substance involves parameters such as the internal energy, the pressure, the specific heat (heat capacity), etc., and such quantities can be derived from models describing the molecular or microsopic behaviour of the substance.

The very term 'statistical mechanics' emphasizes the fact that we must use *statistical* methods to relate the bulk properties of matter to the behaviour of its individual particles. Even a small specimen of matter will contain millions of atoms and we have to calculate the *average* properties of this large number of particles without having information about specific individual particles. We shall also see that these statistical methods can be applied to photons, elastic waves in solids and wave functions as well as to molecules, atoms, electrons, etc.; for our purpose we shall regard all of these as 'particles'.

In recent years much work has been done to extend the methods of statistical mechanics to deal with *non*-equilibrium situations, a step similar to extending classical thermodynamics to irreversible processes. In this book we shall consider only equilibrium statistical mechanics. We shall deal with solids which are more or less perfect crystals and with models of a gas approximating to an ideal gas. The more difficult case of the liquid phase will also be discussed.

2.2 MACROSTATES AND MICROSTATES

Our starting point is to take a specimen of matter—gas, solid or liquid—and regard it as an assembly of particles. So we consider a system comprising an assembly of N identical particles in a volume V. The thermodynamic state of the system is also known as the *macrostate*. This description of the system will contain enough information to enable the thermodynamic state to be clearly defined. We now make a definite decision to limit our system to an *isolated* system, in which case the total internal energy U, as well as N and V, are fixed constants. This means that the macrostate will be defined by:

(a) the fixed amount of the substance, determined by N,
(b) the fixed volume V, since no work is to be done on the system,
(c) the fixed internal energy U, since no energy enters or leaves the system, and
(d) the nature of the system, that is, whether it is a piece of pure aluminium, a given volume of hydrogen, etc.

The (N, U, V) macrostate so described can enables us to obtain, by the methods of thermodynamics, a description as, say, an (N, P, T) macrostate which might be more suitable for our particular substance.

We now proceed to consider what is meant by a *microstate* of the assembly. If we adopt the quantum mechanical approach the wave function ψ_{w} for the whole assembly involves the wave functions of all the N particles. We shall see later (section 5.1) that ψ_{w} contains the product

$$\psi_1 \psi_2 \ldots \psi_N$$

of the wave functions of the individual particles.

The quantum state defined by ψ_{w} describes a microstate of the whole assembly because it gives us details of the state of each of the N particles. The macrostate (N, U, V) will have a large number Ω of microstates corresponding to it. This leads us to the fundamental postulate of statistical mechanics which states that

All possible microstates of an isolated assembly are equally probable.   (2.1)

There is no general proof of this assumption but, as we shall see later, it is justified by the correctness of the various results which follow from it. (For certain assemblies it has been formally proved.)

2.3 AN ASSEMBLY OF DISTINGUISHABLE PARTICLES

To discuss what is meant by an assembly of distinguishable particles we consider an assembly of identical real atoms and ask whether these atoms can be distinguished or 'labelled'. If we take a lattice solid the *position* of each lattice site 'labels' the atom occupying that particular site. We emphasize that there is no question of an atom itself being labelled—the atoms are all identical and indistinguishable—but the labelling is done by the actual geometry of the array of sites. A typical atom is then on site number 5931 (say) and it is the location of site 5931 which labels this atom in much the same way that, for example, the description D15 'labels' a person sitting in row D, seat 15 in a vast crowded auditorium.

If, on the other hand, we have a *gaseous* assembly of identical molecules then quantum mechanics tells us that the molecules are indistinguishable as well as identical. Such assemblies will be dealt with later (Chapters 4 and 5). For the present we return to discuss our assembly of distinguishable particles, which, in some standard texts, are called *localized* particles for the reasons given above.

We next affirm that our assembly consists of *quasi-independent* or *weakly interacting* particles. What we mean is that the particles making up our assembly must interact sufficiently with one another for them all to be in thermal equilibrium. So, since they are not entirely independent we say that they are quasi-independent. There are interactions between them which are sufficient to ensure that they are in thermal equilibrium; they are therefore 'weakly interacting' which means the same as 'quasi-independent'.

In our assembly we have N weakly interacting distinguishable particles each of which has one of its permissible energies $\varepsilon_1, \varepsilon_2, \ldots, \varepsilon_j, \ldots$. The total internal energy is fixed at U and the particles occupy a fixed volume V. Suppose that in any particular distribution we have

$$
\begin{array}{lll}
n_1 & \text{particles with an energy} & \varepsilon_1 \\
n_2 & \text{particles with an energy} & \varepsilon_2 \\
\vdots & & \vdots \\
n_j & \text{particles with an energy} & \varepsilon_j \\
\vdots & & \vdots \\
\text{etc.} & &
\end{array}
\tag{2.2}
$$

Then we must have

$$
\sum_j n_j = N
\tag{2.3}
$$

and

$$
\sum_j n_j \, \varepsilon_j = U
\tag{2.4}
$$

which are two 'restrictive' conditions expressing the constancy of N and U. The second condition (2.4) is a consequence of the fact that the weak interaction energies between the particles may be ignored.

The number of microstates of the assembly corresponding to the distribution (2.2) is the number of ways we can put N distinguishable particles into boxes labelled

$\varepsilon_1, \varepsilon_2, \ldots, \varepsilon_j, \ldots$ so that there are $n_1, n_2, \ldots, n_j, \ldots$ particles in each box. The number of ways of doing this is

$$t(n) = \frac{N!}{n_1! n_2! \ldots n_j! \ldots} \tag{2.5}$$

(see Appendix 2). The total number, Ω, of microstates will be the sum of the terms $t(n)$ for all possible sets of the numbers $n_1, n_2, \ldots, n_j, \ldots$, that is

$$\Omega = \sum t(n) = \sum \frac{N!}{n_1! n_2! \ldots n_j! \ldots}. \tag{2.6}$$

To illustrate these ideas let us consider a very simple assembly consisting of three distinguishable partcles A,B and C. We assume that each of the three particles can assume one of the following energy levels:

$$\varepsilon_0 = 0, \ \varepsilon_1 = \varepsilon, \ \varepsilon_2 = 2\varepsilon, \ \ldots, \ \varepsilon_j = j\varepsilon \quad \text{etc.}$$

where each value of j is an integer.

Let us fix the internal energy U to be 3ε. and let there be

$$
\begin{array}{lll}
n_0 & \text{particles with an energy} & \varepsilon_0 = 0 \\
n_1 & \text{particles with an energy} & \varepsilon_1 = \varepsilon \\
n_2 & \text{particles with an energy} & \varepsilon_2 = 2\varepsilon \\
n_3 & \text{particles with an energy} & \varepsilon_3 = 3\varepsilon
\end{array}
$$

Then

$$U = 3\varepsilon = \sum_j n_j \varepsilon_j \tag{2.7}$$

where n_j is the number of particles each with energy $\varepsilon_j = j\varepsilon$ and where $j = 0, 1, 2$ or 3.

We now consider how this energy U may be divided among three particles. There are three possibilities or *distributions*, I, II, and III, thus:

$$
\begin{array}{llll}
\text{I} & \varepsilon, & \varepsilon, & \varepsilon \\
\text{II} & 3\varepsilon, & 0, & 0 \\
\text{III} & 2\varepsilon, & \varepsilon, & 0
\end{array}
$$

For distribution I, $n_0 = n_2 = n_3 = 0$, $n_1 = 3$ and so, from equation (2.5), the corresponding number of microstates is

$$t(n) = \frac{3!}{3! 0! 0! 0!} = 1$$

putting 0! equal to unity as is usual in this work. For distribution II, $n_0 = 2$, $n_1 = n_2 = 0$, $n_3 = 1$ and the number of microstates is

$$t(n) = \frac{3!}{2!1!0!0!} = 3.$$

Finally for distribution III, $n_0 = n_1 = n_2 = 1$, $n_3 = 0$ and the number of microstates is

$$t(n) = \frac{3!}{1!1!1!0!} = 6.$$

The various possible microstates are shown in Table 2.1

Table 2.1.

A	ε	3ε	0	0	2ε	2ε	ε	ε	0	0
B	ε	0	3ε	0	ε	0	2ε	0	2ε	ε
C	ε	0	0	3ε	0	ε	0	2ε	ε	2ε
	I		II				III			

We see that there are 10 columns representing the 10 ways in which the energy 3ε may be shared out among the three distinguishable particles. Each column represents a microstate and we have $\Omega = 10$ microstates, all equally probable, for this simple assembly.

If we look again at Table 2.1 we see that the 10 microstates can be grouped into three distributions represented by the three sets of ns in columns I, II, and III of Table 2.2.

Table 2.2.

n_0	0	2	1
n_1	3	0	1
n_2	0	0	1
n_3	0	1	0
	I	II	III

Since all microstates are equally probable, we see from Table 2.1 that the probabilities of observing the assembly in the distributions I, II or III are in the ratio 1:3:6. The most probable distribution is III since it will occur in 60% of the cases.

It is instructive to extend the above method to consider an assembly of $N = 6$ particles each with a possible energy $0, \varepsilon, 2\varepsilon$, etc. and put $U = 6\varepsilon$. Let there be

$$n_0 \quad \text{particles with an energy} \quad \varepsilon_0 \ = \ 0$$
$$n_1 \quad \text{particles with an energy} \quad \varepsilon_1 \ = \ \varepsilon$$
$$n_2 \quad \text{particles with an energy} \quad \varepsilon_2 \ = \ 2\varepsilon$$

$$n_6 \quad \text{particles with an energy} \quad \varepsilon_6 \ = \ 6\varepsilon$$

It is found that there are 11 distributions as shown in the first column of Table 2.3; the corresponding numbers of microstates, $t(n)$, are given in the second column.

To illustrate the way of evaluating $t(n)$ we shall consider distributions I and II. For distribution I, $n_0 = 5, n_6 = 1$ and all the other ns are zero. So, from equation (2.5)

$$t(n) = \frac{6!}{5!1!0!0!0!0!0!} = 6.$$

For distribution II, $n_0 = 4$, $n_1 = 1$, $n_5 = 1$ and all the other ns are zero. So

$$t(n) = \frac{6!}{4!1!1!0!0!0!0!} = 30.$$

Table 2.3.

	Distribution	Number of microstates, $t(n)$
I	$6\varepsilon, 0, 0, 0, 0, 0$	6
II	$5\varepsilon, \varepsilon, 0, 0, 0, 0$	30
III	$4\varepsilon, 2\varepsilon, 0, 0, 0, 0$	30
IV	$4\varepsilon, \varepsilon, \varepsilon, 0, 0, 0$	60
V	$3\varepsilon, 3\varepsilon, 0, 0, 0, 0$	15
VI	$3\varepsilon, 2\varepsilon, \varepsilon, 0, 0, 0$	120
VII	$3\varepsilon, \varepsilon, \varepsilon, \varepsilon, 0, 0$	60
VIII	$2\varepsilon, 2\varepsilon, 2\varepsilon, 0, 0, 0$	20
IX	$2\varepsilon, 2\varepsilon, \varepsilon, \varepsilon, 0, 0$	90
X	$2\varepsilon, \varepsilon, \varepsilon, \varepsilon, \varepsilon, 0$	30
XI	$\varepsilon, \varepsilon, \varepsilon, \varepsilon, \varepsilon, \varepsilon$	1

In the same way the other values of $t(n)$ can be calculated and this is left as an exercise for the reader.

The total number of microstates is

$$\Omega = \sum t(n) \tag{2.8}$$

were $\sum$ is the summation over all the distributions. In this case $\Omega = 462$.

As we see from Table 2.3 one distribution occurs 120 times out of a total of 462. For this distribution $t(n)$ will have a maximum value of $t_{\text{max}} = 120$; it is the 'most probable' distribution. From this reasoning we are gradually groping towards a more general result for an assembly, namely, that as N increases more and more it turns out that the most probable distribution (corresponding to t_{max}) stands out increasingly. It is found that only the largest term t_{max} in Ω makes any effective contribution to Ω. As we shall see later we shall really be interested in $\ln \Omega$ rather that Ω itself. So if we write (2.8) as $\Omega = At_{\text{max}}$, then

$$\ln \Omega = \ln t_{\text{max}} + \ln A$$

and our problem is to find how good an approximation $\ln t_{\text{max}}$ is to $\ln \Omega$.

To do this we find the number of microstates, $t(n)$, corresponding to the various distributions for assemblies of $N = 2, 3, \dots, 9, 10$ particles each with a total energy of $U = N\varepsilon$. (It is rather difficult to do this for larger values of N but it is possible to proceed as far as $N = 10$ without being too discouraged!) We pick out t_{max} in each case.

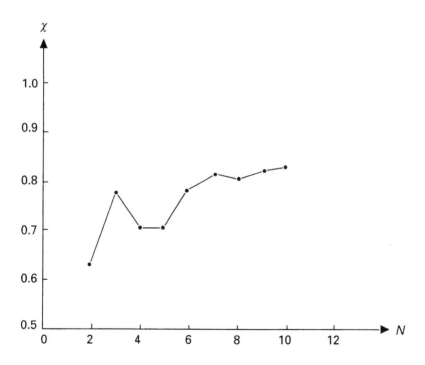

Fig. 2.1. The variation of χ with N.

The results for these different values of N are shown in Table 2.4.

Table 2.4.

N	Total number (Ω) of microstates	$t_{\max}$	$\chi = \frac{\ln t_{\max}}{\ln \Omega}$
2	3	2	0.6310
3	10	6	0.7782
4	35	12	0.6990
5	126	30	0.7034
6	462	120	0.7802
7	1 716	420	0.8108
8	6 435	1 120	0.8005
9	24 310	3 780	0.8158
10	92 378	12 600	0.8256

In the last column the ratio $\chi = \frac{\ln t_{\max}}{\ln \Omega}$ is shown. These values of χ show that, even for these trivially small values of N, $\ln t_{\max}$ is a reasonably good approximation to $\ln \Omega$. The graph of Fig. 2.1 is a plot of χ against N and shows an upward trend as N increases. It is therefore reasonable to assume that as N assumes values much larger than those in Table 2.4, the value of χ tends more and more to the value unity. In practice, in a thermodynamic assembly, N is of the order of 10^{24}, so that we can be reasonably sure that $\chi \to 1$ in this case. This means that, when N is sufficiently large, only the greatest term in Ω makes any effective contribution to Ω, so that (2.8) becomes

$$\Omega \simeq t_{\max} \tag{2.9}$$

when N is sufficiently large.

For further reading to justify this assertion, the reader is referred to Rushbrooke (1949), Trevena (1964) and Offenbacher (1965).

2.4 THE BOLTZMANN DISTRIBUTION

Since, as we have already seen in the previous section, it is only the greatest term on the r.h.s. of (2.6) which makes any significant contribution to Ω, we can drop the summation sign and write

$$\Omega = t_{\max} = \frac{N!}{n_1! n_2! \ldots n_j! \ldots} \tag{2.10}$$

provided the ns in the term on the r.h.s. in (2.10) are those which make this term a maximum. We need to find the values of these ns subject, of course, to our two restrictive conditions (2.3) and (2.4). It is easier to deal with $\ln \Omega$ rather than Ω and we proceed to find the ns which make $\ln \Omega$ a maximum using the method of undetermined multipliers. The procedure is as follows.

For any small changes $\delta n_1, \delta n_2$, etc. in n_1, n_2, etc. there will be a corresponding change in $\ln \Omega$ given by

$$\delta \ln \Omega = \frac{\partial \ln \Omega}{\partial n_1} \delta n_1 + \frac{\partial \ln \Omega}{\partial n_2} \delta n_2 + \dots .$$

If the ns have values which make $\ln \Omega$ a maximum then $\delta \ln \Omega = 0$ for all small changes δn, that is,

$$\frac{\partial \ln \Omega}{\partial n_1} \delta n_1 + \frac{\partial \ln \Omega}{\partial n_2} \delta n_2 + \dots = 0. \tag{2.11}$$

Since N and U are both constant then, from (2.3) and (2.4),

$$\sum_j \delta n_j = 0 \tag{2.12}$$

and

$$\sum_j \varepsilon_j \delta n_j = 0. \tag{2.13}$$

We now multiply (2.12) by α and (2.13) by β and add to (2.11) to obtain, as the condition for the maximum number of microstates,

$$\sum_j \left[\frac{\partial \ln \Omega}{\partial n_j} + \alpha + \beta \varepsilon_j \right] \delta n_j = 0. \tag{2.14}$$

We may regard (2.12) and (2.13) as giving δn_1 and δn_2 in terms of the other independent quantities $\delta n_3, \delta n_4, \dots$ etc.

If we choose α and β so that

$$\left. \begin{array}{c} \dfrac{\partial \ln \Omega}{\partial n_1} + \alpha + \beta \varepsilon_1 = 0 \\[4mm] \text{and} \\[4mm] \dfrac{\partial \ln \Omega}{\partial n_2} + \alpha + \beta \varepsilon_2 = 0 \end{array} \right\} \tag{2.15}$$

then (2.14) becomes

$$\sum_{j>2} \left[\frac{\partial \ln \Omega}{\partial n_j} + \alpha + \beta \varepsilon_j \right] \delta n_j = 0. \tag{2.16}$$

Since all the δn_j in (2.16) are independent this means that for all $j > 2$

$$\frac{\partial \ln \Omega}{\partial n_j} + \alpha + \beta \varepsilon_j = 0. \tag{2.17}$$

Thus, from (2.15) and (2.17) the condition for the maximum number of microstates is

$$\frac{\partial \ln \Omega}{\partial n_j} + \alpha + \beta \varepsilon_j = 0 \tag{2.18}$$

for all j.

(α and β are 'undetermined multipliers' but their values can be determined, as we shall see in Chapter 3.)

For all physically meaningful assemblies the values of N and all the ns are very large. For such large numbers we can use Stirling's theorem (Appendix 1) and write

$$\ln N! \simeq N \ln N - N$$

and $\ln n_j! \simeq n_j \ln n_j - n_j$ for all j.

Thus, from (2.10), we get

$$\ln \Omega = \ln \frac{N!}{n_1! n_2! \ldots n_j! \ldots}$$

$$= \ln N! - \sum_j \ln n_j!$$

$$= \ln N! - \sum_j (n_j \ln n_j - n_j).$$

Thus

$$\frac{\partial \ln \Omega}{\partial n_j} = -\frac{\partial}{\partial n_j}(n_j \ln n_j - n_j)$$

$$= -\ln n_j.$$

Substituting in (2.18) we get

$$-\ln n_j + \alpha + \beta \varepsilon_j = 0.$$

Therefore

$$n_j = e^\alpha e^{\beta \varepsilon_j} = A e^{\beta \varepsilon_j} \tag{2.19}$$

where $A = e^\alpha$. This result in (2.19) is the Boltzmann distribution.

Let us be quite clear as to the meaning of the values of n_j in (2.19). They are the values which correspond to the maximum term on the r.h.s. of (2.6). We shall call this set of ns the *equilibrium* distribution because there are overwhelmingly more microstates corresponding to this set of ns than for all other possible sets of the ns taken together. Since all microstates are equally probable the assembly is vastly more likely to be found in a microstate corresponding to the distribution given by (2.19) than in any other microstate. So we can justifiably refer to the ns given by (2.19) as the most probable or equilibrium distribution.

We note also that though the values of the ns in (2.19) give this most probable or equilibrium distribution, there will be an exceedingly small chance that the assembly will assume a different set of ns. We describe such a situation as a fluctuation from the most probable distribution. However, as has been discussed by Schroedinger (1948) and Wilks (1961), the number of such fluctuations is negligibly small and we may take (2.19) as giving the equilibrium distribution.

On this last point it may make matters clearer if we refer back to the simple assembly of $N = 3$ summarized in Tables 2.1 and 2.2. The number of microstates for the distributions I, II and III are 1,3 and 6 respectively. The 'equilibrium' set of ns corresponds to III, and two 'fluctuations' I and II are, theoretically, possible departures from III.

2.5 ENTROPY AND THE NUMBER OF MICROSTATES

We now seek some equation or relation which will bridge the gap between the equilibrium macroscopic thermodynamic description of an isolated assembly and that of the microscopic.

Historically, Boltzmann showed in 1872 that the entropy S of a gas could be related to the probability of the gas being in a particular state as given by the distribution function $f(v_x, v_y, v_z)$ of its molecular velocities. A further step forward was made by Planck when he extended these ideas to include not only gases but all thermodynamic assemblies. This latter step was possible because Planck's quantum hypothesis had introduced the concept of separate quantum energy states; such discrete energies are much easier to handle than continuously distributed ones.

For an isolated assembly, thermodynamics tells us that the state of equilibrium is one in which the entropy reaches a maximum value. On the other hand, statistical mechanics tells us that the equilibrium state is the most probable since it has the maximum number of microstates. So we can expect some relation between the entropy S and the probability of a state as given by the number of microstates, Ω. As a result of the work of Boltzmann and Planck the relationship

$$S = k_B \ln \Omega \qquad (2.20)$$

was proposed. This is the 'bridge equation' which we need to link the macroscopic and microscopic descriptions of our assembly. (In another field of physics we have the 'bridge equation' $\varepsilon = h\nu$ connecting the particle and wave descriptions of matter.)

Equation (2.20) is really the foundation of statistical mechanics.

Let us now consider two independent assemblies 1 and 2 at the same temperature and pressure. Since entropy is an extensive property (see, for example, Chapter 2 of Finn's *Thermal physics*) the total entropy S is the sum of the entropies of the individual assemblies. Thus

$$S = S_1 + S_2. \tag{2.21}$$

If Ω_1 and Ω_2 are the corresponding numbers of microstates, the total number of microstates for both taken together is

$$\Omega = \Omega_1 \Omega_2 \tag{2.22}$$

since each of the Ω_1 microstates can occur with each of the Ω_2.

So we have

$$S = k_B \ln \Omega = k_B \ln \Omega_1 \Omega_2 = k_B \ln \Omega_1 + k_B \ln \Omega_2$$
$$= S_1 + S_2.$$

Thus equations (2.20), (2.21) and (2.22) are consistent and it is not difficult to show that the logarithm relation (2.20) is the *only* relation for which (2.21) and (2.22) are satisfied.

The numerical value of k_B in (2.20) must be chosen so that the thermodynamic value S of the entropy agrees with its statistical value $k_B \ln \Omega$. We shall see in section 6.6 that k_B turns out to be the Boltzmann constant.

We gain further insight into the concept of entropy if we consider an assembly at the absolute zero $T = 0$. As $T \to 0$ all the N particles in our assembly fall into their ground state, each of the same energy. For N particles, each having the same energy, there is only one microstate (see column I in Table 2.1). So at $T = 0$, $\Omega = 1$, $\ln \Omega = 0$ and $S = 0$. We refer to this condition of the assembly as one of perfect order, since the state of each particle is known uniquely. As T increases more energy becomes available to be shared among the particles, whose individual energies now become unequal; so Ω becomes greater than 1 (see columns II and III in Table 2.1) and S also increases. It is now no longer possible to know the energy state of each particle since this may be different in the various microstates that can occur. We say that the assembly becomes more disordered as the number of possible microstates increases. So Ω, and therefore $S = k_B \ln \Omega$ also, is a measure of this disorder.

One final word about entropy. We shall take equation (2.20) as our definition of entropy rather than define it in the usual thermodynamic way, where a change in S is defined by $dS = dQ_r/T$ (see Finn, 1989, Chapter 5). Later we shall see that both definitions are consistent (see section 6.6).

2.6 A FINAL SUMMARY

For our assembly of N distinguishable particles we have introduced three terms: macrostate, microstate and distribution.

The *macrostate* is described by (N, U, V) together with the nature of the substance concerned.

The *microstate* tells us which, and how many, particles are in each energy state.

The *distribution* lies between the macrostate and microstate. It tells us how many particles are in each energy state but not which ones they are; this latter information is given by the microstate. For example, for $N = 3$, the distributions are shown in Table 2.2 while the more detailed microstates are listed in Table 2.1; there are 3 distributions but 10 microstates.

We have also introduced the relation $S = k_B \ln \Omega$ (equation (2.20)) and the 'averaging' postulate (2.1) as the basis of statistical mechanics. To justify this we can do no better than quote from Rushbrooke (1949).

> 'For the present we must regard the validity of [equation (2.20)] as ultimately resting on the agreement of theoretical results deduced from it with experimental data.'

He goes on to say that the postulate (2.1)

> 'underlies the whole of statistical mechanics and, like [equation (2.20)] must, for the present, be regarded as justified *a posteriori*, by the success of the theory based on it.'

3

The Boltzmann distribution and related topics

In the last chapter we derived the result $n_j = e^{\alpha} e^{\beta \varepsilon_j}$ (equation (2.19)), which is the Boltzmann distribution for an assembly of localized or distinguishable weakly interacting identical particles. As we saw, the obvious example of such an assembly is a crystalline solid in which the particles occupy the lattice 'sites'. In the above equation the ε_js depend on the volume to which each particle is confined; this is a small 'cell' of volume V/N where N is the total number of particles in our assembly of volume V.

We also encountered the equation $S = k_B \ln \Omega$ (equation (2.20)), relating the entropy to the number of microstates. In these two results in equations (2.19) and (2.20) we have introduced three constants α, β and k_B and we shall now proceed to discuss their significance.

3.1 THE CONSTANT α AND THE PARTITION FUNCTION

Since

$$n_j = e^{\alpha} e^{\beta \varepsilon_j} \tag{3.1}$$

then

$$N = \sum_j n_j = e^{\alpha} \sum_j e^{\beta \varepsilon_j}$$

which gives us

$$e^\alpha = \frac{N}{\sum\limits_j e^{\beta \varepsilon_j}}. \tag{3.2}$$

This result gives us α, or at least e^α, in statistical terms but we shall have to wait until later to find out the thermodynamic and physical significance of α (see section 5.5).

Substituting for e^α in (3.1) we get

$$n_j = N \frac{e^{\beta \varepsilon_j}}{\sum\limits_j e^{\beta \varepsilon_j}}. \tag{3.3}$$

The sum $\sum\limits_j e^{\beta \varepsilon_j}$ in the denominator of (3.2) is important and is known as the *partition function* (p.f.) for a particle of the assembly. From (3.3) we see that $n_1, n_2, n_3, \ldots$ etc. are respectively proportional to the so-called Boltzmann factors $e^{\beta \varepsilon_1}, e^{\beta \varepsilon_2}, e^{\beta \varepsilon_3}, \ldots$ etc. So the p.f. tells us how the particles are 'partitioned' or split up between the different energy levels.

We write the p.f. for a particle of the assembly as

$$Z = \sum_j e^{\beta \varepsilon_j} \tag{3.4}$$

so that (3.3) becomes

$$n_j = \frac{N e^{\beta \varepsilon_j}}{Z} \tag{3.3a}$$

and (3.2) can be written as

$$\alpha = \ln N - \ln Z. \tag{3.5}$$

3.2 THE CONSTANTS β AND k_B

We first convince ourselves that β is related to temperature by considering two assemblies of distinguishable particles in thermal equilibrium but otherwise isolated from the rest of the universe. The two assemblies 1 and 2 of N and N' particles will then be at the same temperature, T. We consider a given distribution for assembly 1 to be described by

$$n_j \text{ particles with energy } \varepsilon_j \text{ for all } j$$

and one for assembly 2 by

$$n_r' \text{ particles with energy } \varepsilon_r' \text{ for all } r.$$

Following and extending the method of section 2.4 the total number of microstates Ω is

$$\Omega = \Omega_1 \Omega_2$$

$$= \frac{N!}{n_1! n_2! \ldots n_j! \ldots} \times \frac{N'!}{n_1'! n_2'! \ldots n_r'! \ldots}$$

(cf. equation (2.22)) and

$$\ln \Omega = N \ln N - \sum_j n_j \ln n_j + N' \ln N' - \sum_r n_r' \ln n_r'.$$

We require the values of the ns and $n's$ which make $\ln \Omega$ a maximum subject to our restrictive conditions which now take the form

$$
\left.
\begin{aligned}
\sum_j n_j &= N & \text{(a)} \\
\sum_r n_r' &= N' & \text{(b)} \\
\sum_j n_j \varepsilon_j + \sum_r n_r' \varepsilon_r' &= U & \text{(c)}
\end{aligned}
\right\} \tag{3.6}
$$

(cf. equations (2.3) and (2.4)).

Proceeding as in section 2.4 we have

$$
\left.
\begin{aligned}
\alpha_1 \sum_j \delta n_j &= 0 & \text{(a)} \\
\alpha_2 \sum_r \delta n_r' &= 0 & \text{(b)} \\
\beta \left(\sum_j \varepsilon_j \delta n_j + \sum_r \varepsilon_r' \delta n_r' \right) &= 0 & \text{(c)}
\end{aligned}
\right\} \tag{3.7}
$$

and we get, finally,

$$\sum_j (-\ln n_j + \alpha_1 + \beta \varepsilon_j) \delta n_j + \sum_r (-\ln n_r' + \alpha_2 + \beta \varepsilon_r') \delta n_r' = 0$$

which, to hold for all possible values of δn_j and $\delta n_r'$, gives us

$$n_j = e^{\alpha_1} e^{\beta \varepsilon_j}, \quad n_r' = e^{\alpha_2} e^{\beta \varepsilon_r'}.$$

This result shows that the equilibrium distribution for both assemblies 1 and 2 is of the Boltzmann kind. Each distribution has its own particular value of α, which is not surprising in view of the two *separate* equations (3.7a,b) for particle conservation. On the other hand, the equation (3.7c) involving β is a single equation involving the energy conservation of both assemblies and is a result of the thermal contact (that is, energy interchange) between them. So we would really expect β to be the same for both distributions and so it turns out to be. Since both β and T are the same for our two assemblies, this suggests that β is some sort of measure of the temperature, T.

We now delve further into this relation between β and T by considering a single assembly as in section 2.4. However, we now take an assembly which is no longer isolated since we shall gradually add a quantity of heat δQ to it. This means that in our previous (N, U, V) description of an assembly N and V will both be constant but U will not. Since the εs depend on V they will not change as the heat δQ is added but the ns will adjust in value to keep Ω at a maximum. So

$$\delta Q = \delta U = \delta \sum_j n_j \varepsilon_j = \sum_j \varepsilon_j \delta n_j .$$

The change in $\ln \Omega$ as a result of adding δQ will be

$$\delta \ln \Omega = \sum_j \frac{\partial \ln \Omega}{\partial n_j} \delta n_j$$

$$= -\sum_j (\alpha + \beta \varepsilon_j) \delta n_j \qquad \text{from (2.18)}$$

$$= -\alpha \sum_j \delta n_j - \beta \sum_j \varepsilon_j \delta n_j$$

$$= -\beta \delta Q,$$

since $\sum_j \delta n_j = 0$ for constant N.

So

$$\beta = -\frac{d \ln \Omega}{dQ}. \tag{3.8}$$

From the second law of thermodynamics, in any reversible change, we have

$$dQ = T dS$$

$$= T d(k_B \ln \Omega) \qquad \text{from (2.20)}$$

$$= k_B T d \ln \Omega$$

$$= -k_B T \beta dQ \qquad \text{from (3.8)}.$$

Therefore $\beta = \frac{-1}{k_B T}$; but we have still to show that this k_B is the Boltzmann constant. For the present we shall assume that it is and justify it later when we compare the equation of state of a gas derived from statistical mechanics with the ideal gas equation (see section 6.6).

So from now on we write

$$\beta = \frac{-1}{k_{\mathrm{B}}T}. \tag{3.9}$$

3.3 THE RELATION BETWEEN THE PARTITION FUNCTION AND OTHER THERMODYNAMIC FUNCTIONS

In section 3.1 we emphasized the importance of the partition function. This lies in the fact that it can be used to determine other thermodynamic functions, as we shall now show.

Before doing so let us remind ourselves that we derived the Boltzmann distribution for an assembly described by (U, V, N). Since $\beta = \frac{-1}{k_{\mathrm{B}}T}$ this distribution can be written as

$$n_j = e^\alpha e^{-\varepsilon_j/k_{\mathrm{B}}T}$$

and we therefore see that the distribution (and assembly) may be described by (T, V, N) just as well as by (U, V, N). To use the (T, V, N) description is really more convenient in practice because we are more likely to know T than U for an assembly. With this approach U and other quantities can be obtained in terms of (T, V, N).

At this point we shall gather together the main results which we have so far derived for an assembly of distinguishable (localized) particles. They are

$$\Omega = \frac{N!}{\prod_j n_j!} \tag{a}$$

$$n_j = e^\alpha e^{\beta\varepsilon_j} = \frac{N e^{\beta\varepsilon_j}}{Z} \tag{b}$$

$$N = \sum_j n_j = e^\alpha Z \tag{c}$$

$$U = \sum_j n_j\varepsilon_j \tag{d}$$

$$Z = \sum_j e^{\beta\varepsilon_j} \tag{e}$$

$$S = k_{\mathrm{B}}\ln\Omega \ (= k_{\mathrm{B}}\ln t_{\max})$$
$$\quad = k_{\mathrm{B}}(N\ln N - \sum_j n_j \ln n_j) \tag{f}$$

$$\beta = \frac{-1}{k_{\mathrm{B}}T} \tag{g}$$

$$(3.10)$$

So for the entropy we have

$$S = k_{\mathrm{B}} \left[N \ln N - \sum_j n_j \ln n_j \right]$$

$$= k_{\mathrm{B}} \left[N \ln N - \sum_j n_j (\alpha + \beta \varepsilon_j) \right] \quad \text{from (3.10b)}$$

$$= k_{\mathrm{B}} \left[N \ln N - \alpha N - \beta U \right] \quad \text{from (3.10c,d)}$$

$$= k_{\mathrm{B}} \left[N \ln Z + \frac{U}{k_{\mathrm{B}} T} \right], \tag{3.11}$$

using (3.10c,g).

Defining the Helmholtz free energy F as

$$F = U - TS \tag{3.12}$$

we get at once from (3.11)

$$F = -N k_{\mathrm{B}} T \ln Z \tag{3.13a}$$

$$= -k_{\mathrm{B}} T \ln Z^N \tag{3.13b}$$

$$= -k_{\mathrm{B}} T \ln Z_{\mathrm{W}} \tag{3.13c}$$

where

$$Z_{\mathrm{W}} = Z^N \tag{3.14}$$

can be regarded as the partition function for the assembly of N distinguishable (localized) particles. We shall consider this point further in section 3.5.

Two other useful results from thermodynamics are

$$S = -\left(\frac{\partial F}{\partial T} \right)_V \tag{3.15}$$

and

$$P = -\left(\frac{\partial F}{\partial V} \right)_T, \tag{3.16}$$

(see section 6.6).

Next we derive U in terms of Z. From (3.10b and d) we have

$$U = \sum_j n_j \varepsilon_j = \mathrm{e}^\alpha \sum_j \varepsilon_j \mathrm{e}^{\beta \varepsilon_j}.$$

Using (3.10c and g) we get

$$U = \frac{N}{Z} \sum_j \varepsilon_j e^{-\varepsilon_j/k_\mathrm{B}T}$$

$$= \frac{Nk_\mathrm{B}T^2}{Z} \frac{\partial}{\partial T} \left(\sum_j e^{-\varepsilon_j/k_\mathrm{B}T} \right)$$

$$= \frac{Nk_\mathrm{B}T^2}{Z} \frac{\partial Z}{\partial T}$$

$$= Nk_\mathrm{B}T^2 \frac{\partial}{\partial T} \ln Z. \tag{3.17}$$

3.4 DEGENERACY

In our discussion so far we have assumed that for each of our energy levels ε_j there is only one stationary wave function ψ_j, that is, we have assumed that each of these energy levels is *non-degenerate*. This assumption was implicit in the various results we have derived, such as those summarized in (3.10). However we know from quantum mechanics that it is possible that more than one wave function can exist for a given energy level ε_j; when this is so the energy level is *degenerate* and the number of wave functions g_j corresponding to ε_j is known as the *degeneracy* or *statistical weight* of that level.

For example, let us consider a single isolated atom performing a harmonic vibration with a certain amplitude and therefore a certain energy. It is possible that its axis of vibration can take up a number of different directions, each corresponding to a different wave function but to the same energy.

To proceed let us consider the energy level $j = 1$ with n_1 particles each with energy ε_1. It has g_1 independent wave functions corresponding to it. The first of our n_1 particles can be in any one of g_1 wave states, and so can the second particle; this means that the first and second particles can, together, be in one of g_1^2 states. Continuing this argument we see that our n_1 particles can be in one of $g_1^{n_1}$ states.

If we further extend this argument for all values of j we see that there will be an extra factor of $g_1^{n_1} g_2^{n_2} \ldots g_j^{n_j} \ldots$ in the result of equation (2.10) which now becomes modified to

$$t_{\max} = \frac{N!}{n_1! n_2! \ldots n_j! \ldots} g_1^{n_1} g_2^{n_2} \ldots g_j^{n_j} \ldots \tag{3.18a}$$

$$= \frac{N!}{\prod_j n_j!} \prod_j g_j^{n_j} \tag{3.18b}$$

where the ns correspond to the maximum term $t_{\max}$. If we carry out the same analysis as before (see section 2.4) we find that the partition function becomes modified to

$$Z = \sum_j g_j \mathrm{e}^{-\varepsilon_j/k_\mathrm{B}T}$$

and equation (3.3) now becomes

$$n_j = N \frac{g_j \mathrm{e}^{\beta \varepsilon_j}}{\sum_j g_j \mathrm{e}^{\beta \varepsilon_j}}.$$

The previous equations involving Z, namely (3.11), (3.13) and (3.17), remain formally unchanged provided we use this modified form of Z for the partition function. The modification simply means that a degenerate state is counted g_j times.

3.5 THE PARTITION FUNCTION FOR THE ASSEMBLY

For any one of an assembly of N identical localized particles there is a partition function $Z = \sum_j \mathrm{e}^{\beta \varepsilon_j}$ (equation (3.4)), assuming the energy levels are non-degenerate.

For the assembly as a whole we also define a partition function

$$Z_\mathrm{W} = \sum \mathrm{e}^{\beta E_N}$$

where the various values of E_N represent the possible values of the total energy of the assembly, that is, the sum of the N individual εs. The summation extends over all possible values of E_N.

To see how Z_W is related to Z we first consider a simple assembly of just two localized (distinguishable) particles situated at sites a and b in a lattice; they are therefore 'labelled' by their positions in the lattice. Using equation (3.4) we can write the partition functions of our two particles as

$$Z_\mathrm{a} = \sum_j \mathrm{e}^{\beta \varepsilon_{ja}} = \mathrm{e}^{\beta \varepsilon_{1a}} + \mathrm{e}^{\beta \varepsilon_{2a}} + \mathrm{e}^{\beta \varepsilon_{3a}} + \dots$$

and

$$Z_\mathrm{b} = \sum_j \mathrm{e}^{\beta \varepsilon_{jb}} = \mathrm{e}^{\beta \varepsilon_{1b}} + \mathrm{e}^{\beta \varepsilon_{2b}} + \mathrm{e}^{\beta \varepsilon_{3b}} + \dots$$

Now the various values of the total energy of the assembly are given by

$$E_N = \varepsilon_\mathrm{a} + \varepsilon_\mathrm{b}$$

where ε_a assumes one of the values $\varepsilon_{1a}, \varepsilon_{2a}, \dots$ etc. and likewise ε_b assumes one of the values $\varepsilon_{1b}, \varepsilon_{2b}, \dots$ etc. The partition function Z_W must take into account all possible values of E_N. So

$$Z_W = e^{\beta(\varepsilon_{1a}+\varepsilon_{1b})} + e^{\beta(\varepsilon_{1a}+\varepsilon_{2b})} + e^{\beta(\varepsilon_{1a}+\varepsilon_{3b})} + \cdots$$

$$+ e^{\beta(\varepsilon_{2a}+\varepsilon_{1b})} + e^{\beta(\varepsilon_{2a}+\varepsilon_{2b})} + \cdots$$

$$+ e^{\beta(\varepsilon_{3a}+\varepsilon_{1b})} + \cdots$$

etc.

$$= \sum_j e^{\beta\varepsilon_{ja}} \sum_j e^{\beta\varepsilon_{jb}}$$

$$= Z_a Z_b$$

However, since the particles at a and b are identical then $\varepsilon_{ja} = \varepsilon_{jb} = \varepsilon_j$ for all j and $Z_a = Z_b = Z$. Thus $Z_W = Z^2$. By extending this argument to an assembly of N particles we have the result

$$Z_W = Z^N. \tag{3.14}$$

We emphasize that this result (3.14) holds for an assembly of localized particles, which is the type of assembly to which the Boltzmann distribution applies (section 3.3). Even though the particles are identical we can distinguish between them by virtue of the fact that they occupy different positions in a lattice. We shall see in section 6.6 how the result in (3.14) has to be modified when the particles are not distinguishable.

4

The approach to gases

4.1 INTRODUCTION

Up to the present we have considered assemblies of quasi-independent distinguishable particles which are localized at definite lattice sites. We now discuss the other extreme in which we have a gas of non-localized particles (atoms or molecules) free to move in a container of volume V. The energy levels (states) available to our particles are determined mainly by the extent (that is, geometrical boundaries) of V and these levels are very closely spaced in energy, as we shall see in section 4.5.

The gas particles, no longer 'labelled' by their positions in a lattice, are identical and indistinguishable. To deal properly with this indistinguishability we have to resort to quantum mechanics; this will be done in the next chapter.

We make one other observation. Since the gas particles are assumed to be quasi-independent they are not close enough together to interact with each other, which implies that the density of the gas is low. So our work in this chapter is mainly concerned with a low-density gas.

4.2 THE 'PARTICLE IN A BOX' PROBLEM

Since we are going to consider a gaseous assembly let us first concentrate on just one gas particle free to move in any direction in a cubical box of side a. The origin is taken to be at one corner of the box. Using quantum mechanics the stationary-state wave function $\psi(x, y, z)$ which describes the particle is of the form

$$\psi = A\sin(n_x \pi x/a)\sin(n_y \pi y/a)\sin(n_z \pi z/a) \tag{4.1}$$

where the numbers n_x, n_y and n_z are positive integers $1, 2, 3, \ldots$ etc. This stationary-state form of ψ ensures that the particle is confined within the box since $\psi = 0$ over the surface of the box. It also shows that an integral number n_x, n_y and n_z of half-waves fit into the box in the x, y and z directions. The wavelengths λ of the possible stationary waves are therefore

$$\lambda = \frac{2a}{n_x}, \ \frac{2a}{n_y} \text{ and } \frac{2a}{n_z}$$

in the x, y and z directions.

Since the momentum of the particle is given by $p = h/\lambda$ the components of its momentum are

$$p_x = n_x \ h/2a, \ p_y = n_y \ h/2a, \ p_z = n_z \ h/2a.$$

The resultant momentum is given by

$$\begin{aligned} p_j^2 = p_x^2 + p_y^2 + p_z^2 &= (n_x^2 + n_y^2 + n_z^2)h^2/4a^2 \\ &= n_j^2 h^2/4a^2 \end{aligned}$$

and the corresponding energy

$$\begin{aligned} \varepsilon_j = p_j^2/2m &= h^2(n_x^2 + n_y^2 + n_z^2)/8ma^2 \\ &= n_j^2 h^2/8ma^2. \end{aligned} \tag{4.2}$$

The wave state of the particle is defined by the values of n_x, n_y and n_z but its possible discrete energy levels ε_j correspond to the different possible values of n_j^2 and *not* to the individual values of n_x, n_y and n_z. Various combinations of n_x, n_y and n_z satisfy the equation $n_x^2 + n_y^2 + n_z^2 = n_j^2$ for a given value of n_j; if there are g_j such combinations there will be g_j different wave states corresponding to the same energy ε_j. The energy level or energy state is then said to be degenerate and its degeneracy is g_j.

Since the volume of the cubical box is $V = a^3$ equation (4.2) can be written as

$$\varepsilon_j = n_j^2 h^2/8m \ V^{2/3}. \tag{4.3}$$

Although this result has been derived for a cubical box it holds for a container of volume V of any shape. We emphasize that equation (4.3) shows that the energy depends on the quantum number n_j and on the volume V.

The lowest energy level $j = 1$ is that for which $n_x = n_y = n_z = 1$ and $n_j^2 = n_1^2 = 3$. So $\varepsilon_1 = 3h^2/8m \ V^{2/3}$ and there is only one wave function (that is, one set of quantum numbers n_x, n_y, n_z) in this case. Hence this lowest energy level is non-degenerate and $g_1 = 1$.

One final word. We have so far considered only one particle in our box. If there are several particles and if they are *quasi-independent*, their allowed energies are the same as those of a single particle given by equations (4.2) and (4.3).

4.3 THE DENSITY OF STATES

Let us consider equation (4.2) and write

$$\varepsilon = h^2(n_x^2 + n_y^2 + n_z^2)/8ma^2 = h^2n^2/8ma^2 \tag{4.4}$$

where we now omit the suffix j.

n_x, n_y and n_z are positive integers and by giving them various values we can represent the corresponding values of ε as a series of points forming a simple cubic lattice in $n_x n_y n_z$ space (or n-space) with rectangular axes On_x, On_y and On_z. Rewriting (4.4) we have

$$n_x^2 + n_y^2 + n_z^2 = n^2 = 8ma^2\varepsilon/h^2$$
$$= R^2 \text{ (say)}.$$

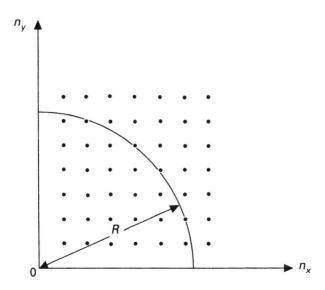

Fig. 4.1. A cross-section, in the $n_x n_y$ plane, through the positive octant of radius R, showing the lattice points which lie in this plane.

So the number of energy levels (states) with energy less than ε is equal to the number of points in n-space in the positive octant of the sphere of radius R (see Fig. 4.1). It is not difficult to see that the number of such points is equal to the volume of this octant as long as the sphere is not too small. (A similar situation is encountered in lattices in crystallography.) Hence the number of levels with energy less than ε is

$$\begin{aligned} N &= \frac{1}{8} \times \frac{4\pi R^3}{3} \\ &= \frac{4\pi}{3}\left(\frac{2ma^2\varepsilon}{h^2}\right)^{3/2}. \end{aligned} \tag{4.5}$$

If we consider a slightly larger octant containing $N + dN$ points corresponding to an energy $\varepsilon + d\varepsilon$ then the number of levels or states in the energy range $d\varepsilon$ is

$$\begin{aligned} dN &= \frac{2\pi a^3}{h^3}(2m)^{3/2}\varepsilon^{1/2}d\varepsilon \\ &= g(\varepsilon)d\varepsilon, \quad \text{(say)}. \end{aligned}$$

So

$$g(\varepsilon)d\varepsilon = \frac{2\pi V}{h^3}(2m)^{3/2}\varepsilon^{1/2}d\varepsilon \tag{4.6}$$

since $V = a^3$.

The function g so defined is the *density of states in energy*. The number of states with energies between ε and $\varepsilon + d\varepsilon$ is $g(\varepsilon)d\varepsilon$.

It is also instructive to derive the result (4.6) as follows. Referring to equation (4.4) the number of such energy levels with ns in the small interval $n \to n + dn$ is

$$g(n)dn = \frac{1}{8} \times 4\pi n^2 dn.$$

The r.h.s. of this equation is the volume, in the positive octant, in our n-space between the spheres of radii n and $n + dn$. To calculate $4\pi n^2 dn$ in terms of ε etc. we use equation (4.3) and write

$$n^2 = \frac{8mV^{2/3}\varepsilon}{h^2} = \alpha\varepsilon \ \text{(say)}.$$

Then $2ndn = \alpha d\varepsilon$ and $n = \alpha^{1/2}\varepsilon^{1/2}$.

So

$$\begin{aligned} 4\pi n^2 dn &= 2\pi\alpha^{3/2}\varepsilon^{1/2}d\varepsilon \\ &= \frac{16\pi V}{h^3}(2m)^{3/2}\varepsilon^{1/2}d\varepsilon. \end{aligned}$$

So $g(n)dn$, which is the same as $g(\varepsilon)d\varepsilon$, is one-eighth of this, which agrees with (4.6).

The result (4.6) for $g(\varepsilon)\mathrm{d}\varepsilon$ was based on the value of ε given by (4.2), which takes into account the translational motion only of a particle of the gas. We have, in fact, neglected the spin of the particle; in other words, the result (4.6) is valid for particles of spin zero. For example, in the case of fermions with a spin quantum number $s = \frac{1}{2}$, there will be two spin states for each translational state and we must therefore multiply the r.h.s. of (4.6) by 2. Fermions will be dealt with explicitly in Chapters 5 and 7.

Furthermore, in addition to spin, we have the possibility of internal vibration and rotation occurring if our particle is a molecule. These matters will be discussed in Chapter 6.

4.4 MORE ABOUT DENSITY OF STATES

We can rewrite (4.1) as

$$\psi = A \sin k_x x \sin k_y y \sin k_z z$$

where $k_x = n_x \pi/a$, $k_y = n_y \pi/a$ and $k_z = n_z \pi/a$.

So we can use the ks instead of the ns to specify the possible states of ψ. k_x, k_y and k_z are positive integers and can be considered as the components of a wave vector $\mathbf{k}$ of scalar magnitude k. Thus we can use the idea of k-space in which a point of coordinates (k_x, k_y, k_z) represents the k-state of ψ. All such points form a simple cubic lattice in the positive octant of k-space, the lattice spacing being π/a for consecutive values of the n_x, n_y and n_z. In other words each unit cube in the lattice has a volume of $(\pi/a)^3$.

We now consider the number of k-states whose values of k lie between k and $k + \mathrm{d}k$. This number will be equal to the volume in the positive octant of the spherical shell of radius k and thickness $\mathrm{d}k$ divided by $(\pi/a)^3$, the volume of the unit cube. We write this number as $g(k)\mathrm{d}k$. Therefore

$$g(k)\mathrm{d}k = \frac{1}{8} \times \frac{4\pi k^2 \mathrm{d}k}{(\pi/a)^3}$$

$$= \frac{V}{(2\pi)^3} \times 4\pi k^2 \mathrm{d}k. \tag{4.7}$$

The function $g(k)$ is the *density of states in* k. In fact the quantities $g(k)\mathrm{d}k$, $g(\varepsilon)\mathrm{d}\varepsilon$ and $g(n)\mathrm{d}n$ all have the same dimensions and are pure numbers. As before we can include the spin in $g(k)\mathrm{d}k$ by multiplying the r.h.s. of (4.7) by an appropriate factor when necessary.

Since $\varepsilon = n^2 h^2/8ma^2$ and $k^2 = n^2\pi^2/a^2$ then

$$\varepsilon = k^2 h^2/8m\pi^2 = k^2\hbar^2/2m, \tag{4.8}$$

where $\hbar = h/2\pi$. Using this $\varepsilon - k$ relation (the dispersion relation) it is possible to transform (4.7) to obtain the form of $g(\varepsilon)\mathrm{d}\varepsilon$ in (4.6). This is left as a simple exercise for the reader.

4.5 SOME USEFUL CALCULATIONS

The energy levels in a gaslike model are extremely closely spaced. We can show this by using equation (4.3) in the form

$$\varepsilon = h^2 n^2 / 8 m V^{2/3}.$$

The energy difference $\Delta\varepsilon$ between adjacent levels $n = n$ and $n = (n + 1)$ will be of the order

$$\Delta\varepsilon \sim h^2 / m V^{2/3}. \tag{4.9}$$

Let us now do some calculations based on a sample of helium (^{4}He) gas at 290 K contained in a cubical box of volume $V = 10^{-3}\,m^3$, that is, $a = 10$ cm. Helium, of course, is a monatomic gas. If we substitute for h, m and V in (4.9) we obtain $\Delta\varepsilon \sim 6.6 \times 10^{-39}\,J$. At $T = 290\,K$, $k_B T = 4.0 \times 10^{-21}\,J$. So $\Delta\varepsilon$ is very small compared with $k_B T$ and it follows that there will be a very large number of energy levels available for energies less than $k_B T$. Since $\Delta\varepsilon$ is so small we are almost always justified in regarding the energy levels in our gaslike model as a continuum which means that sums over energy levels can be replaced by integrations.

We next use the above example of helium to show that there are many more levels than atoms. In doing so we are following the procedure given by Wilks (1961). The velocities of the helium atoms can be measured by a rotating drum method such as that described in Tabor's (1979) book *Gases, liquids and solids*, 2nd edition, Chapter 4. It is found that many have velocities as high as 2.5×10^3 ms^{-1} corresponding to an energy of $2 \times 10^{-20}\,J$. So the energy levels ε range from the lowest values to at least this value of $\varepsilon = 2 \times 10^{-20}\,J$. Putting this value of ε into (4.5) gives the number of such levels in a 10 cm cube as $N = 6.2 \times 10^{28}$. The number of atoms per cubic centimetre in helium gas at atmospheric pressure is 2.5×10^{19} so that in our 10 cm cube there are 2.5×10^{22} atoms.

The ratio

$$\frac{\text{number of available levels}}{\text{number of atoms}} = \frac{6.2 \times 10^{28}}{2.5 \times 10^{22}} \sim 10^6$$

shows that there are at least of the order of 10^6 times more energy levels than atoms. For gases for which m is larger this ratio will be even greater.

To sum up, the above calculations have shown two things for our gaslike model. Firstly, we have seen that the number of available energy levels is very large indeed and these levels are very close together. Secondly, for a real 'chemical' gas under normal conditions, the number of energy levels far exceeds the number of atoms (or molecules).

4.6 THE DISTRIBUTION FUNCTION FOR A REAL CHEMICAL GAS UNDER NORMAL CONDITIONS

As we have seen, for a real gas under normal conditions, the number of available energy levels (or states) far exceeds the number of particles (atoms or molecules). This implies that the occupation number per level is usually 0, sometimes 1 and hardly ever greater than 1. How do we deal with this state of affairs?

The method adopted is to group the energy levels into bundles. A typical bundle, the kth bundle, is shown schematically in Fig. 4.2. It contains g_k levels which are occupied by a total of N_k particles and $g_k \gg N_k$. The mean energy of the levels in the bundle is denoted by ε_k. In doing this we must ensure that we satisfy two requirements, namely:

(1) N_k must be a large enough number so that Stirling's theorem can be applied.
(2) The number of levels g_k must not be too large so that the mean energy ε_k is sufficiently close to that of each level in the bundle. Since these levels are closely spaced in energy we can be reassured on this point.

We emphasize that the precise values of all our g_ks are not important provided the two above conditions are satisfied.

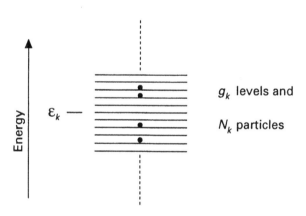

Fig. 4.2. The kth bundle of energy levels (or states).

A distribution of the gaslike assembly may now be described by specifying the values of N_k for each ε_k, that is, stating the number of particles in each bundle. Note how this differs from our treatment of distinguishable (localized) particles in Chapters 2 and 3 where the distribution was given by stating the number (n_j) of particles in each *individual* energy level (ε_j).

The next task is to evaluate the number of microstates. To do this we first assume that the particles are distinguishable (labelled). Then, referring to Fig. 4.2, the first particle can be inserted into one of the g_k levels in g_k ways. The second, similarly, can be inserted in g_k ways, so that the first two can therefore be inserted in g_k^2 ways. Continuing this argument we see that the N_k particles can be inserted in $g_k^{N_k}$ ways. Of course this number $g_k^{N_k}$ will include the cases in which several, or indeed all, of the particles occupy one level but, since $g_k \gg N_k$, the number of such cases is very small so that the number $g_k^{N_k}$ is very close to the answer we require.

If we refer to Fig. 4.2, the four dots, which initially represented *labelled* particles, can be arranged among themselves in 4! ways. In general, for N_k particles, this number would be $N_k!$ ways. (The 4! and $N_k!$ do not include permutations in the same energy level but the probability of having more than one particle in a level is negligible.) If we remove the labels from the particles the 4! and $N_k!$ arrangements are both simply replaced by *one* arrangement. It follows that the result $g_k^{N_k}$ for distinguishable particles overestimates, by a factor $N_k!$, the number of microstates for particles which are indistinguishable. Hence, for indistinguishable particles, the number of microstates for the kth bundle is

$$\Omega_k = g_k^{N_k}/N_k!$$

If we consider all the bundles, the total number of microstates is the product of all the Ω_ks since each of the microstates in any one bundle can occur with each microstate in each other bundle. Hence the total number of microstates is

$$\Omega = \prod_k \Omega_k$$
$$= \prod_k (g_k^{N_k}/N_k!) \tag{4.10}$$

Our two restrictive conditions now are

$$N = \sum_k N_k$$

and

$$U = \sum_k N_k \varepsilon_k.$$

As before we maximize $\ln \Omega$ by applying the method of undetermined multipliers (see section 2.4) The condition for the most probable state is

$$\frac{\partial \ln \Omega}{\partial N_k} + \alpha + \beta \varepsilon_k = 0. \tag{4.11}$$

From (4.10),

$$\ln \Omega = \sum_k [N_k \ln g_k - N_k \ln N_k + N_k]. \tag{4.12}$$

Therefore

$$\frac{\partial \ln \Omega}{\partial N_k} = \ln(g_k/N_k).$$

Substituting this in (4.11) and rearranging then gives the most probable value as

$$N_k = g_k e^{\alpha} e^{\beta \varepsilon_k} = A g_k e^{\beta \varepsilon_k} \tag{4.13a}$$

where A and α are found from the equation $N = \sum_k N_k$; in fact, $A = N \Big/ \sum_k g_k e^{\beta \varepsilon_k}$.

So

$$N_k = N g_k e^{\beta \varepsilon_k} \Big/ \sum_k g_k e^{\beta \varepsilon_k} \tag{4.13b}$$

on substituting for A.

The equilibrium number of particles per level (state) in the bundle k is

$$f_k = \frac{N_k}{g_k} = \exp(\alpha + \beta \varepsilon_k). \tag{4.14}$$

f_k is called the distribution function.

The results in equations (4.10), (4.13) and (4.14) are the classical or Maxwell–Boltzmann results; (4.14) is a Maxwell–Boltzmann distribution function.

In the heading to this section we described our gas as a 'chemical' gas. By this we mean one whose macroscopic thermodynamic behaviour is adequately described by its pressure, volume and temperature.

Finally, we emphasize that the treatment in this section is valid only if $g_k \gg N_k$ for all k, that is, the energy states are very thinly populated, usually with 0 or 1 particle in each. For this reason the gas is often called a *dilute* gas; in practice real chemical gases at low density and high temperature (such as room temperature) are in this category.

5

A more detailed treatment of gases

In Chapter 4 we discussed the treatment of a gaseous assembly of particles. In particular we derived the distribution function for such an assembly at high temperatures and low densities (equation (4.14)). In order to extend our theory to deal with more general conditions, that is, all temperatures, it is necessary for us to consider quantum mechanics and how it deals with indistinguishability.

5.1 SYMMETRIC AND ANTISYMMETRIC WAVE FUNCTIONS

We start by considering a simple assembly of two identical quasi-independent indistinguisable particles with energies ε_1 and ε_2. Let their wave functions be $\psi_1(q_1)$ and $\psi_2(q_2)$ where each q represents all the coordinates (space and spin coordinates) of the particles concerned. The quantity $|\psi_1(q_1)|^2$ represents the probability of finding particle 1 at q_1 and $|\psi_2(q_2)|^2$ that of finding particle 2 at q_2. As far as the *assembly* of two particles is concerned the total energy is $\varepsilon_1 + \varepsilon_2$ and quantum mechanics shows us that the Schroedinger equation for the assembly is satisfied by a wave function $\psi(q_1, q_2)$ which is of the product form

$$\psi_1(q_1)\psi_2(q_2) \tag{5.1}$$

or

$$\psi_1(q_2)\psi_2(q_1)$$

or any linear combination of these two.

Then $|\psi(q_1, q_2)|^2$ gives the probability of finding one particle at q_1 and the other at q_2. If we permute the particles, that is, interchange the qs, this probability becomes $\psi|(q_2, q_1)|^2$ which must be the same as $\psi|(q_1, q_2)|^2$ since the particles are indistinguishable. So we have

$$|\psi(q_1, q_2)|^2 = |\psi(q_2, q_1)|^2.$$

Hence

$$\psi(q_1, q_2) = \pm\psi(q_2, q_1). \tag{5.2}$$

The possibility of a negative or positive sign in this last equation leads us to consider two types of particle. If we first take the negative sign then

$$\psi(q_1, q_2) = -\psi(q_2, q_1). \tag{5.3}$$

Stated in words this equation tells us that interchanging q_1 and q_2 results in a change in sign of the wave function which we therefore say is *antisymmetric*. Particles which behave like this are called *fermions*.

The wave function for the assembly of two particles is the linear combination

$$\psi_{\mathrm{w}} = \psi(q_1, q_2) = \psi_1(q_1)\psi_2(q_2) - \psi_1(q_2)\psi_2(q_1). \tag{5.4}$$

If we interchange q_1 and q_2 then the r.h.s. of (5.4) changes sign and so condition (5.3) is satisfied. Furthermore, a little thought will show that the linear combination (5.4) is the only one that will satisfy condition (5.3).

We now generalize this theory to an assembly of N particles with a total energy of $\varepsilon_1 + \varepsilon_2 + \ldots + \varepsilon_N$. The product wave function

$$\psi_1(q_1)\psi_2(q_2)\ldots\psi_N(q_N) \tag{5.5}$$

satisfies the Schroedinger equation for the assembly of N particles (cf. equation 5.1). So will any of the $N!$ product wave functions of type (5.5) obtained by permuting the qs in the N brackets in (5.5). The antisymmetric wave function ψ_{w} for the assembly is now a generalization of (5.4) and is the linear combination of the $N!$ wave functions of type (5.5).

So we have

$$\begin{aligned}
\psi_{\mathrm{w}} = {} & \psi_1(q_1)\psi_2(q_2)\ldots\psi_N(q_N) \\
& -\psi_1(q_2)\psi_2(q_1)\ldots\psi_N(q_N) \\
& +\text{etc.} \\
& -\text{etc.}
\end{aligned} \tag{5.6}$$

On the r.h.s. of (5.6) there are $N!$ product terms, half of them with positive signs and the other half with negative signs. In fact ψ_{w} can be written as a determinant thus:

$$\psi_{\mathrm{W}} = \begin{vmatrix} \psi_1(q_1) & \psi_1(q_2) & \dots & \psi_1(q_N) \\ \psi_2(q_1) & \psi_2(q_2) & \dots & \psi_2(q_N) \\ \vdots & \vdots & & \vdots \\ \psi_N(q_1) & \psi_N(q_2) & \dots & \psi_N(q_N) \end{vmatrix} \qquad (5.7)$$

The wave function ψ_{W} in (5.6) and (5.7) has the required property of antisymmetry since interchanging any two of the qs—say q_1 and q_2—results in an interchange of two columns of the determinant which leads only to a change in sign of the determinant. Thus the probability $|\psi_{\mathrm{W}}|^2$ remains unchanged.

Furthermore, if any two of the particles has the same wave function then ψ_{W} disappears. For example if $\psi_1 = \psi_2$ then two rows of the determinant will be the same and it vanishes identically. Hence if we have an assembly of particles (fermions) described by an antisymmetric wave function, not more than one particle may be in the same state, that is, occupy the same non-degenerate energy state (level). This is the Pauli exclusion principle.

Next let us consider (5.2) with the positive sign, that is

$$\psi(q_1, q_2) = \psi(q_2, q_1).$$

This tells us that interchanging q_1 and q_2 does not change the sign of the wave function for the assembly of two particles and so this wave function is symmetric. Such particles are called *bosons*. The symmetric wave function for the assembly is now obtained by replacing the negative sign on the r.h.s. of (5.4) by a positive sign, that is

$$\psi_{\mathrm{W}} = \psi_1(q_1)\psi_2(q_2) + \psi_1(q_2)\psi_2(q_1).$$

In the same way, the symmetric wave function for an assembly of N bosons is the sum of all the $N!$ product wave functions in (5.6) but with all the signs positive. In this case *any number* of particles may be in the same state, that is, energy level.

We next ask: which particles, in fact, turn out to be bosons or fermions? It is observed experimentally that particles with integral spin $(0, 1, 2, \dots$ etc.) are bosons; examples are photons (spin 1) and ^{4}He atoms (spin 0). On the other hand particles with half-integral spin $(\frac{1}{2}, \frac{3}{2}, \dots$ etc.) are fermions, examples being electrons and ^{3}He atoms, each with spin $\frac{1}{2}$.

5.2 THE NUMBER OF MICROSTATES FOR FERMIONS AND BOSONS

The microstate is represented by the relevant ψ_{W} (symmetric or antisymmetric) of the previous section. Each ψ_{W} is described by the individual particle states labelled $1, 2, \dots, N$. The counting of these microstates is based on the bundle treatment of section 4.6. Because of the difference in the occupation of individual states for fermions and bosons, we shall have to treat the two cases separately. For both, the diagram of Fig. 4.2 is appropriate. We have a bundle of g_k energy states (levels) of mean energy ε_k and we require to put N_k particles (fermions or bosons) into these g_k states. As in section 4.6 the distribution is described by stating how many particles there are in each bundle of states.

We now consider fermions and bosons in turn.

(a) Fermions

Because of the Pauli exclusion principle each energy state can contain either 0 or 1 particle. We consider the kth bundle. So the problem is this: in how many ways can the N_k states, each containing a fermion, be chosen from a total of g_k states? The answer is $g_k C_{N_k}$ or

$$\frac{g_k!}{N_k!(g_k - N_k)!}$$

for the number of microstates. This implies that $N_k \leqslant g_k$, that is, the number of fermions cannot exceed the number of states.

For *all* the bundles the total number of microstates is therefore

$$\Omega_{\mathrm{FD}} = \prod_k \frac{g_k!}{N_k!(g_k - N_k)!}. \tag{5.8}$$

The statistics of fermions, based on (5.8), are known as the Fermi–Dirac (FD) statistics.

(b) Bosons

In this case there is no restriction on the number of bosons that may occupy any given particle state.

To evaluate the number of microstates we first consider 10 bosons which are to be inserted into 4 states. If we represent the 10 bosons as a row of 10 dots, thus

. . . .

we can parcel them into 4 lots by inserting 3 (that is, 4 minus 1) lines. Hence

. . | . | . . . . | . . .

is one possibility.

The total number of such possibilities is found as follows. We have a total of 13 symbols (dots and lines) in a row and we require the number of ways of choosing 3 positions for the lines out of a total of 13; this number is $13C_3$, that is,

$$\frac{13!}{3!10!}.$$

More generally, for our kth bundle, we can consider N_k particles (dots) in a row to be split up into g_k lots by inserting $(g_k - 1)$ lines. So we have a total of

$(N_k + g_k - 1)$ symbols and we must choose $(g_k - 1)$ positions for our lines; the number of such choices is

$$\frac{(N_k + g_k - 1)!}{N_k!(g_k - 1)!}.$$

For all the bundles we therefore have

$$\Omega_{\text{BE}} = \prod_k \frac{(N_k + g_k - 1)!}{N_k!(g_k - 1)!} \tag{5.9a}$$

for our total number of microstates.

Since $g_k \gg 1$ we can use g_k instead of $(g_k - 1)$ and (5.9a) can be written as

$$\Omega_{\text{BE}} = \prod_k \frac{(N_k + g_k)!}{N_k! g_k!}. \tag{5.9b}$$

The equations (5.9a,b) give the number of microstates for a boson gas and this assembly of bosons is treated by the Bose–Einstein (BE) statistics.

5.3 THE DISTRIBUTION FUNCTIONS FOR FERMIONS AND BOSONS

To derive the distribution functions for fermions and bosons we use the same method as in section 4.6 and maximize $\ln \Omega$ subject to the two restrictive conditions $N = \sum_k N_k$ and $U = \sum_k N_k \varepsilon_k$. The condition for the most probable state is

$$\frac{\partial \ln \Omega}{\partial N_k} + \alpha + \beta \varepsilon_k = 0. \tag{5.10}$$

(a) Fermions

From (5.8)

$$\ln \Omega_{\text{FD}} = \sum_k [g_k \ln g_k - N_k \ln N_k - (g_k - N_k) \ln(g_k - N_k)]$$

using Stirling's theorem.

Thus

$$\frac{\partial \ln \Omega_{\text{FD}}}{\partial N_k} = \ln \left(\frac{g_k - N_k}{N_k} \right).$$

Substituting in (5.10) gives

$$\ln \left(\frac{g_k - N_k}{N_k} \right) + \alpha + \beta \varepsilon_k = 0.$$

Therefore

$$N_k = \frac{g_k}{e^{-\alpha - \beta \varepsilon_k} + 1}$$

or

$$f_k = \frac{N_k}{g_k} = \frac{1}{e^{-\alpha - \beta \varepsilon_k} + 1}. \tag{5.11}$$

This is the Fermi–Dirac distribution function, f_k, and it is the (equilibrium) average number of particles per energy state. f_k does not depend on the (arbitrary) value chosen for g_k, which is a satisfactory feature; it is a function of the energy only.

(b) Bosons

From (5.9b)

$$\ln \Omega_{\mathrm{BE}} = \sum_k [(N_k + g_k) \ln(N_k + g_k) - N_k \ln N_k - g_k \ln g_k].$$

Therefore

$$\frac{\partial \ln \Omega_{\mathrm{BE}}}{\partial N_k} = \ln \left(\frac{N_k + g_k}{N_k} \right).$$

Substituting in (5.10) we get

$$\ln \left(\frac{N_k + g_k}{N_k} \right) + \alpha + \beta \varepsilon_k = 0.$$

Therefore

$$N_k = \frac{g_k}{e^{-\alpha - \beta \varepsilon_k} - 1}$$

or

$$f_k = \frac{N_k}{g_k} = \frac{1}{e^{-\alpha - \beta \varepsilon_k} - 1}. \tag{5.12}$$

This is the Bose–Einstein distribution function.

5.4 THE DILUTE GAS: THE CLASSICAL LIMIT

We now consider the case of a 'dilute' gas, by which we mean one for which $N_k \ll g_k$ for all values of k; in other words the energy states are very thinly populated. We shall see that under these conditions equations (5.8) and (5.9) reduce to equation (4.14) for the 'high-temperature, low-density' gas we discussed in section 4.6.

We first deal with fermions and consider equation (5.8). We have

$$\frac{g_k!}{N_k!(g_k - N_k)!} = \frac{g_k(g_k - 1)\ldots(g_k - N_k + 1)}{N_k!}$$

where, on the r.h.s., we have divided both numerator and denominator by $(g_k - N_k)!$. There are N_k factors on the top of the r.h.s. of this equation and since $g_k \gg N_k$ each of these factors is approximately equal to, but a little less than, g_k. So the r.h.s. is approximately equal to $g_k^{N_k}/N_k!$ and (5.8) becomes

$$\Omega_{\mathrm{FD}} \simeq \prod_k \frac{g_k^{N_k}}{N_k!}$$

for a dilute fermion gas.

Similarly, considering (5.9b) for bosons we have

$$\frac{(N_k + g_k)!}{N_k!g_k!} = \frac{(N_k + g_k)(N_k + g_k - 1)\ldots(g_k + 1)}{N_k!}.$$

Since $g_k \gg N_k$, each of the N_k factors on the top of the r.h.s. is approximately equal to, but somewhat greater than, g_k.

Hence (5.9b) becomes

$$\Omega_{\mathrm{BE}} \simeq \prod_k \frac{g_k^{N_k}}{N_k!}$$

for a dilute boson gas.

So for this dilute case both the FD and BE results tend (one from above, one from below) to the same classical result

$$\Omega = \prod_k \frac{g_k^{N_k}}{N_k!}$$

which is the Maxwell–Boltzmann (MB) result we considered in Chapter 4 (see equation (4.10)).

This classical result may also be deduced as follows. From (5.11) for fermions we have

$$\frac{N_k}{g_k} = \frac{1}{\exp(-\alpha + \varepsilon_k/k_{\mathrm{B}}T) + 1} \tag{5.13}$$

where we have put $\beta = -1/k_{\mathrm{B}}T$.

If T is large the number of available energy states becomes very large and the particles are spread over this larger number of states. So now $g_k/N_k \gg 1$. Hence

$$\frac{g_k}{N_k} = \exp(-\alpha + \varepsilon_k/k_BT) + 1$$

is a large number and the '1' on the r.h.s. may be ignored. Hence (5.13) becomes

$$N_k = g_k \exp(\alpha - \varepsilon_k/k_BT) \tag{5.14}$$

which is the MB or classical result (see equation 4.13a).

By the same argument the BE result for bosons also reduces to this classical result at high temperatures.

5.5 CLOSED AND OPEN SYSTEMS; α AND THE CHEMICAL POTENTIAL

Consider a system (that is, a sample of matter) consisting of an assembly of particles. The system is separated from its surroundings by a wall or boundary across which an exchange of energy and matter may occur between them.

In a so-called *closed* system there will be exchange of energy only and no exchange of matter; in other words there is a fixed amount of matter in the system. When two such systems are brought into thermal contact there will be no net flow of energy (heat) between them if the parameter $\beta (= -1/k_BT)$ is the same for both.

On the other hand, for two *open* systems (assemblies), it is possible to have an exchange of matter across the boundary separating them. An example is that of two phases of the same substance 'open' to each other at the same temperature such as a liquid and its vapour. In this case there is no *net* transfer of particles from one assembly to the other.

We have already mentioned the role of the parameter β (in section 3.2) and we now consider how the parameter α enters into the picture. We return to the behaviour of a gaslike assembly obeying either the Fermi–Dirac or Bose–Einstein statistics. We have, from (5.11) and (5.12),

$$f_k = \frac{N_k}{g_k} = \frac{1}{\exp(-\alpha + \varepsilon_k/k_BT) \pm 1}$$

(+ for F.D., − for B.E.) The parameter α appearing explicitly in this equation cannot be readily eliminated, as in the case of localized particles. It is determined by the condition $\sum_k N_k = N$, that is, that the total number of particles in the assembly is kept fixed. We shall rewrite the equation as

$$f_k = \frac{N_k}{g_k} = \frac{1}{\exp[(\varepsilon_k - \mu)/k_BT] \pm 1}$$

where $\mu = \alpha k_B T$ is known as the *chemical potential*. Clearly μ has the dimensions of an energy and can be shown to be the Gibbs free energy per particle, provided only one type of particle is present (that is, so long as we have one substance only to deal with); see Appendix 4.

We have seen that α is determined by the condition that the total number of particles in the assembly is kept fixed. Suppose that we have two systems which are two phases of the same substance 'open' to each other at the same temperature such as a liquid (1) and its vapour (2). There will be no *net* transfer of particles from one assembly to the other provided both assemblies have the same value of α (or μ), that is, $\alpha_1 = \alpha_2$ or $\mu_1 = \mu_2$. Of course since the two assemblies are at the same temperature they will also have the same value of β.

To sum up: for two open assemblies of the same substance to be in equilibrium the values of α and β must be the same for both of them; in other words the chemical potentials and temperatures for both are the same.

5.6 A FINAL SUMMARY OF THE THREE DISTRIBUTION FUNCTIONS

We end this chapter by summarizing the three distribution functions for a gaseous assembly:

$$f_k = \frac{1}{e^{-\alpha} e^{\varepsilon_k / k_B T} \pm 1}$$
$$(+\text{FD}, -\text{BE}),$$

$$f_k = e^{\alpha} e^{-\varepsilon_k / k_B T}$$
$$(\text{MB}).$$

6

The Maxwell–Boltzmann distribution for monatomic and diatomic gases

PART I MONATOMIC GASES

6.1 INTRODUCTION

We start by reminding ourselves of the basic assumptions underlying the concept of a perfect or ideal gas. Such a gas is assumed to consist of identical molecules of zero size exerting no forces on one another. They are uniformly distributed in density and perform completely random motion (that is, randomness of both velocity and direction) with all collisions being perfectly elastic. This gas obeys the ideal gas equation

$$PV = qRT \tag{6.1}$$

if there are q moles of the gas present and where R is the gas constant ($R = 8.3144$ J mol^{-1}K^{-1}).

On the basis of these assumptions, Maxwell, in 1859, derived his famous result for the distribution of molecular speeds, as mentioned in Chapter 1.

In most of this book we shall consider only perfect gases, but we shall extend the definition of a perfect gas to include a (real) gas whose particles are quasi-independent (weakly interacting); we shall also regard the particles not as 'points' of zero size but as identical tiny spheres whose diameter is small compared with their average distance apart, which is certainly the situation at low densities. For such a gas the deviations from equation (6.1) are small enough to be neglected.

6.2 A BRIEF SUMMARY OF THE DISTRIBUTION FOR MOLECULAR SPEEDS

The perfect gas we are considerering in Part 1 of this chapter will be assumed to be both monatomic and zero-spin. We take a gas consisting of N molecules, each of mass m, and concentrate on the number dN with velocities between v and $v+dv$, irrespective of their direction of motion. Since there are virtually no intermolecular forces the only energy possessed by each molecule is its kinetic energy $\frac{1}{2}mv^2$. If v has components v_x, v_y and v_z we can consider velocity space as in Fig. 6.1. For any molecule the speed is given by the distance of the corresponding point in velocity space from the origin.

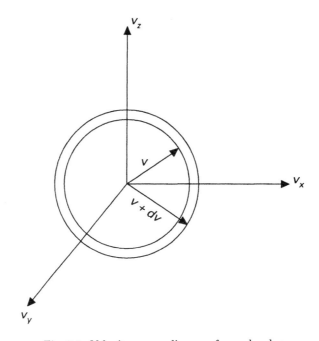

Fig. 6.1. Velocity space diagram for molecules.

From Maxwell's theory the number of molecules dN whose values of v lie inside the spherical shell of radii v and $v + dv$ is given by

$$\frac{1}{N}\frac{dN}{dv} = 4\pi v^2 \left(\frac{m}{2\pi k_\mathrm{B}T}\right)^{3/2} \exp\left(\frac{-mv^2}{2k_\mathrm{B}T}\right) \tag{6.2}$$

which we can rewrite as

$$\phi(v) = \frac{1}{N}\frac{\mathrm{d}N}{\mathrm{d}v} = 4\pi v^2 \left(\frac{a}{\pi}\right)^{3/2} e^{-av^2} \tag{6.3}$$

where $a = m/2k_{\mathrm{B}}T$.

The quantity $\phi(v)$ is plotted against the (scalar) value of v in Fig. 6.2. The function $\phi(v)$ is the Maxwell–Boltzmann distribution function for molecular speeds.

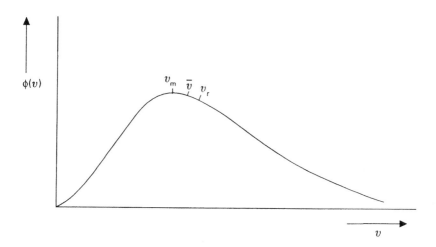

Fig. 6.2. The Maxwell–Boltzmann distribution function for molecular speeds.

There are three values of v which are of particular interest:

(a) The most probable speed v_m corresponding to the maximum of the graph of Fig. 6.2. Putting $\mathrm{d}\phi/\mathrm{d}v = 0$ we readily find from (6.3) that

$$v_m = (2k_{\mathrm{B}}T/m)^{1/2}.$$

(b) The mean (or average) speed $\bar{v}$ is given by

$$\bar{v} = \int_{v=0}^{v=\infty} v\frac{\mathrm{d}N}{N} = \int_0^\infty v\phi(v)\mathrm{d}v = \left(\frac{8k_{\mathrm{B}}T}{\pi m}\right)^{1/2}.$$

(c) The root mean square (r.m.s.) speed $v_r = (\overline{v^2})^{1/2}$ is given by

$$v_r = (\overline{v^2})^{1/2} = \left(\int_{v=0}^{v=\infty} v^2\frac{\mathrm{d}N}{N}\right)^{1/2} = \left(\int_0^\infty v^2\phi(v)\mathrm{d}v\right)^{1/2}$$
$$= (3k_{\mathrm{B}}T/m)^{1/2}.$$

In evaluating $\bar{v}$ and v_r we have used the results

$$\int_0^\infty x^3 e^{-ax^2} dx = \frac{1}{2a^2}$$

and

$$\int_0^\infty x^4 e^{-ax^2} dx = \frac{3}{8a^2} \left(\frac{\pi}{a}\right)^{1/2}$$

respectively.

We find that $\bar{v} = 1.128 v_m$ and $v_r = 1.225 v_m$ and the rough positions of the three speeds are shown in Fig. 6.2.

From the average square speed $v_r^2 = 3k_B T/m$ we find that the average kinetic energy per molecule is

$$\frac{1}{2} m v_r^2 = 3k_B T/2$$

in agreement with the result from the kinetic theory of gases.

In general, for the speed v of a molecule we have

$$v^2 = v_x^2 + v_y^2 + v_z^2.$$

Multiplying by $\frac{1}{2}m$ and averaging the squared speeds we have

$$\frac{1}{2}m\overline{v^2} = \frac{1}{2}m\overline{v_x^2} + \frac{1}{2}m\overline{v_y^2} + \frac{1}{2}m\overline{v_z^2} = \frac{3}{2}k_B T.$$

So the total translational energy is made up of three parts and these three must be equal because of the random directions of the molecules (that is, $\overline{v_x^2} = \overline{v_y^2} = \overline{v_z^2}$). Hence the average kinetic energy for each of the x, y and z directions is $k_B T/2$. These three translational motions in these three directions are called degrees of freedom. All this is in agreement with the law of equipartition of energy, which states that the average energy per molecule per degree of freedom is equal to $k_B T/2$.

The distribution for molecular speeds was derived by Maxwell. Another way of deriving it is to use the methods of statistical mechanics based essentially on Boltzmann's work; this will be done in section 6.5. For these reasons the result in equation (6.2) is now usually referred to as the Maxwell–Boltzmann speed distribution.

6.3 THE PARTITION FUNCTION FOR A MAXWELL–BOLTZMANN (MB) GAS

Since the Maxwell–Boltzmann (dilute) limit $N_k << g_k$ holds (see section 5.4) we have, from equation (5.14), that

$$N_k = g_k e^{\alpha} e^{-\varepsilon_k / k_B T}.$$

We shall examine this limit more closely in the next section.

To make progress we write

$$N = \sum_k N_k$$

$$= e^\alpha \sum_k g_k e^{-\varepsilon_k/k_B T}$$

$$= e^\alpha Z \tag{6.4}$$

where

$$Z = \sum_k g_k e^{-\varepsilon_k/k_B T} \tag{6.5}$$

is the partition function per particle of our MB gas. A little thought will convince us that the sum $\sum_k g_k e^{-\varepsilon_k/k_B T}$ over all the groups k is the same as the sum $\sum_j e^{-\varepsilon_j/k_B T}$ (equation (3.4)) over all the individual energy states j. It is just that we sum over all the states, in different ways, in the two cases.

The summation in (6.5) can be replaced by an integration, the number of states between ε and $\varepsilon + d\varepsilon$ being given by equation (4.6). Thus

$$Z = \int_0^\infty g(\varepsilon) e^{-\varepsilon/k_B T} d\varepsilon$$

$$= 2\pi V \left(\frac{2m}{h^2}\right)^{3/2} \int_0^\infty \varepsilon^{1/2} e^{-\lambda \varepsilon} d\varepsilon$$

where $\lambda = 1/k_B T$.

Now $\int_0^\infty \varepsilon^{1/2} e^{-\lambda \varepsilon} d\varepsilon = \pi^{1/2}/2\lambda^{3/2}$. Hence

$$Z = V \left(\frac{2\pi m k_B T}{h^2}\right)^{3/2} \tag{6.6}$$

We emphasise that Z is the partition function for the translational motion of one particle of our MB gas, at temperature T confined to a volume V.

6.4 A LOOK AT THE VALIDITY OF THE MAXWELL–BOLTZMANN (I.E. THE DILUTE OR CLASSICAL) LIMIT

Consider the Maxwell–Boltzmann distribution written as

$$\frac{N_k}{g_k} = e^\alpha e^{-\varepsilon_k/k_B T},$$

(cf. equation (4.14)).

The condition $N_k/g_k \ll 1$ for all k will always hold if the first exponential is very much less than unity. So the condition $e^\alpha \ll 1$ is necessary for the validity of the classical or MB statistics. We therefore ask the question: is $e^\alpha \ll 1$?

Now, from (6.4) and (6.6)

$$e^\alpha = \frac{N}{Z} = \frac{N}{V}\left(\frac{h^2}{2\pi m k_{\rm B} T}\right)^{3/2}. \tag{6.7}$$

As an example of a monatomic zero-spin gas we consider ^{4}He, since, being a substance with a small m, it will have a comparatively large e^α. To estimate e^α we take our ^{4}He to be at one atmosphere pressure. Its boiling point is 4.2 K so let us take a temperature $T = 4.5$ K corresponding to the gas phase. Substituting the appropriate values into (6.7) gives $e^\alpha \sim 0.1$. So, in this case, e^α is small enough for MB statistics to be a good approximation. For higher values of m and T (e.g. at room temperatures), e^α will be even smaller and so the MB limit can be justifiably used for real chemical gases.

6.5 DERIVATION OF THE MB SPEED DISTRIBUTION

The MB distribution of (4.13a) can be rewritten as

$$N(\varepsilon)d\varepsilon = g(\varepsilon)d\varepsilon e^\alpha e^{-\varepsilon/k_{\rm B}T} \tag{6.8}$$

where $N(\varepsilon)d\varepsilon$ is the number of molecules with energies between ε and $\varepsilon+d\varepsilon$ and $g(\varepsilon)d\varepsilon$ the number of states in the same energy range. Using equations (4.6) and (6.7) for $g(\varepsilon)d\varepsilon$ and e^α equation (6.8) becomes

$$N(\varepsilon)d\varepsilon = \frac{2N}{\pi^{1/2}(k_{\rm B}T)^{3/2}}\varepsilon^{1/2}e^{-\varepsilon/k_{\rm B}T}\,d\varepsilon. \tag{6.9}$$

Now $\varepsilon = \frac{1}{2}mv^2$ for a molecule with speed v; thus $d\varepsilon = mvdv$ and we can readily transform (6.9) to give the number of molecules with velocities between v and $v+dv$ as

$$N(v)dv = 4\pi N\left(\frac{m}{2\pi k_{\rm B}T}\right)^{3/2}v^2 e^{-mv^2/2k_{\rm B}T}\,dv \tag{6.10}$$

which is the same result as equations (6.2) and (6.3) obtained by Maxwell as the law of distribution of molecular speeds. Maxwell, however, did not derive it in the way given in this section but on the basis of more classical ideas.

The law also holds for gases with internal degrees of freedom (see Part II of this chapter).

6.6 THE RELATION WITH THERMODYNAMICS

We start by bringing together some of our previous results.

$$\Omega = \prod_k \frac{g_k^{N_k}}{N_k!} \qquad \text{from equation (4.10)}$$

$$N_k = g_k e^\alpha e^{\beta \varepsilon_k} \qquad \text{from equation (4.13a)}$$

$$\therefore \quad \ln(g_k/N_k) = -\alpha - \beta \varepsilon_k$$

$$N = \sum_k N_k = e^\alpha \sum_k g_k e^{\beta \varepsilon_k} = Z e^\alpha \qquad \text{from equation (6.5)}$$

$$\therefore \quad \alpha = \ln N - \ln Z$$

$$U = \sum_k N_k \varepsilon_k = \sum_k g_k \varepsilon_k e^\alpha e^{\beta \varepsilon_k}$$

$$Z = \sum_k g_k e^{\beta \varepsilon_k} = V \left(\frac{2\pi m k_B T}{h^2} \right)^{3/2} \qquad \text{from equation (6.6)}$$

$$\ln N! = N \ln N - N$$

$$\beta = -1/k_B T.$$

We now use the above results to relate the entropy S, the Helmholtz free energy $F = U - TS$ and the internal energy U to the partition function Z. The algebra, though a little tedious, is straightforward.

$$\ln \Omega = \sum_k [N_k \ln(g_k/N_k)] + N$$
$$= -\alpha N - \beta U + N$$
$$= N \ln Z - \ln N! - \beta U.$$

The entropy is therefore

$$S = k_B \ln \Omega$$
$$= k_B [N \ln Z - \ln N! + U/k_B T] \qquad (6.11)$$

So

$$F = U - TS$$
$$= -k_B T [N \ln Z - \ln N!] \qquad (6.12)$$
$$= -k_B T \ln(Z^N/N!). \qquad (6.13)$$

The quantity $Z^N/N!$ is often called the total or complete partition function Z_W of the assembly. So, writing

$$Z_W = Z^N/N!$$

then

$$F = -k_B T \ln Z_W \tag{6.14}$$

for our assembly of non-localized particles.

Following Chapter 6 of Finn (1986) we have for an infinitesimal change

$$dF = dU - TdS - SdT.$$

Also, since

$$TdS = dU + PdV$$

then

$$dF = -PdV - SdT$$
$$= \left(\frac{\partial F}{\partial V}\right)_T dV + \left(\frac{\partial F}{\partial T}\right)_V dT.$$

Thus, if we know F as a function of V and T, then we can, by comparing coefficients on the r.h.s. of the last two lines, obtain the pressure and entropy. Thus

$$P = -\left(\frac{\partial F}{\partial V}\right)_T$$

and

$$S = -\left(\frac{\partial F}{\partial T}\right)_V.$$

(We met these two results in section 3.3.)

For the pressure, using (6.12),

$$P = -\left(\frac{\partial F}{\partial V}\right)_T = Nk_B T \left(\frac{\partial \ln Z}{\partial V}\right)_T. \tag{6.15a}$$

Using (6.6) $\partial \ln Z/\partial V = 1/V$ and so

$$P = Nk_B T/V$$

or

$$PV = Nk_B T \tag{6.15b}$$

This is the equation of state for our monatomic MB gas of N molecules in our volume V. It must also be identical to the ideal gas law (equation (6.1)), namely

$$PV = qRT$$

where $q = N/N_A$ is the number of moles of gas present and N_A the Avogadro constant. Thus

$$PV = NRT/N_A. \tag{6.16}$$

Comparing (6.15b) and (6.16) gives us $k_B = R/N_A$, the gas constant per molecule which is, of course, the Boltzmann constant. We have therefore verified that our k_B is in fact the Boltzmann constant and so justified our earlier results $\beta = -1/k_B T$ and $S = k_B \ln \Omega$ for our statistical definitions of temperature and entropy (see section 2.5 and section 3.2).

We next consider the internal energy U. We have

$$\frac{U}{N} = \frac{\sum_k g_k \varepsilon_k e^{\beta \varepsilon_k}}{\sum_k g_k e^{\beta \varepsilon_k}}$$

$$= \frac{d}{d\beta} \ln \left(\sum_k g_k e^{\beta \varepsilon_k} \right)$$

$$= \frac{d}{d\beta} (\ln Z)$$

$$= k_B T^2 \frac{d(\ln Z)}{dT}.$$

So

$$U = N k_B T^2 d(\ln Z)/dT. \tag{6.17}$$

(This result also holds for the internal energy of an assembly of *localized* particles for which $U = e^{\alpha} \sum_j \varepsilon_j e^{\beta \varepsilon_j}$ and $N = e^{\alpha} \sum_j e^{\beta \varepsilon_j}$.) Using (6.6), $d(\ln Z)/dT = 3/2T$ and so

$$U = 3N k_B T/2 \tag{6.18}$$

which agrees with the discussion in section 6.2 and also shows that U depends on T only.

Also the heat capacity is

$$C_V = dU/dT = 3N k_B/2. \tag{6.19}$$

Finally we consider the entropy S. We can write (6.12) as

$$F = -N k_B T[\ln Z - \ln N + 1]$$

using Stirling's theorem. Substituting for Z from (6.6) we then differentiate F w.r.t. T keeping V constant; this is left as an exercise for the reader. The result is

$$S = -\left(\frac{\partial F}{\partial T} \right)_V$$

$$= N k_B \left[\frac{3}{2} \ln \left(\frac{2\pi m k_B T}{h^2} \right) - \ln \left(\frac{N}{V} \right) + \frac{5}{2} \right]. \tag{6.20}$$

This well-known result, known as the Sackur–Tetrode equation, gives us the entropy of the gas.

6.7 COMPARISON WITH THE RESULTS FOR LOCALIZED PARTICLES: THE FACTOR $N!$

As we see from equation (6.13), the expression for F contains a factor $N!$; this was absent in the expression for F in the case of localized particles (equation (3.13b)). The same is true for the expressions for S. On the other hand, the results for U and C_V are formally the same for localized and non-localized particles. We may regard the $N!$ as being a 'correction' for the indistinguishability of the non-localized particles.

Consider then our gaseous assembly of N particles. We are interested in their translational motions and we count the distinguishable microstates of the assembly from the positions and velocities of the particles. Suppose first that these particles are hypothetically labelled 1 to N and let us take a flash 'photograph' of the assembly at a given instant, as in Fig. 6.3(a).

As time goes on the particles will present a number of physically identical photographs but with the labelled particles interchanging positions and velocities. One such photograph will be as in Fig. 6.3(b), where the situations (positions and velocities) of particles 1 and 2 have been interchanged, the situations of the other particles being unchanged. Clearly it is possible to have a total of $N!$ photographs, identical to those below, obtained by permuting the labels $1, 2, 3, \ldots, N$.

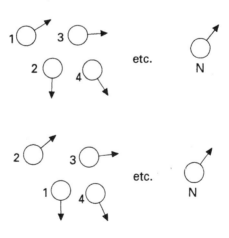

Fig. 6.3a,b.

As Rushbrooke puts it:

> 'Then any particular microstate of the assembly is just one member of a set of $N!$ such states obtained from this one by permuting the labels $1, 2, \ldots, N$ and these $N!$ states of the assembly are mutually distinguishable provided that the labels are attached. Thus all the distinguishable microstates of the hypothetical assembly of labelled particles fall into sets of $N!$ states, the members of any set being mutually distinguishable only on account of the labels artificially attached to the particles. Consequently, if we enumerate the microstates of an assembly of N identical non-localised particles by the expedient of attaching hypothetical labels to the particles, then we count all the truly distinguishable states of the assembly $N!$ times and the true number of microstates is the apparent number (for an assembly of labelled particles) divided by $N!$.'

It follows that indistinguishability results in the appearance of a factor $N!$ in the partition function for the assembly. So instead of the result $Z_{\mathrm{w}} = Z^N$ for an assembly of localized particles we get $Z_{\mathrm{w}} = Z^N/N!$ for one of non-localized particles.

To be strictly correct, we must make a further observation. The quantity $N!$ is the number of ways in which N distinguishable particles can be put into N *different* quantum states. However, if some of the states are the same, then the number of distinct microstates is not $N!$ but $N!/\prod_j n_j!$ (see equation (2.5)). But, as we saw in section 4.6, since in this case the number of states greatly exceeds the number of gas particles, then the number of states occupied by more than one particle is negligibly small; that is, each n_j is either 0 or 1 so that $N!/\prod_j n_j!$ effectively reduces to $N!$.

PART II DIATOMIC GASES

We now consider a diatomic gas, assumed to obey MB statistics. We concentrate first on a single diatomic molecule in which the two atoms are bound together by an attractive force along the axis of the molecule.

6.8 THE PARTITION FUNCTION FOR A DIATOMIC MOLECULE

We can consider the energy of a diatomic molecule as being made up of four independent contributions. These are:

ε_t, the translational energy due to the translational motion of the centre of mass of the molecule.

ε_r, the rotational energy due to the rotation of the two atoms about the centre of mass of the molecule.

ε_ν, the vibrational motion of the two atoms along the axis of the molecule.
ε_e, the energy of the atomic electrons.

It is assumed that the four contributions to the total energy of our molecule can vary independently and we write

$$\varepsilon_i = \varepsilon_t + \varepsilon_r + \varepsilon_\nu + \varepsilon_e \tag{6.21}$$

for the ith eigenstate of our molecule.

Because of the independence of the four energy contributions it means that the partition function Z for the molecule factorizes; we can see this as follows.

The partition function of the diatomic molecule is

$$Z = \sum_{(\text{states})} \exp(\beta\varepsilon_i)$$

$$= \sum_{(\text{states})} \exp(\beta\varepsilon_t)\exp(\beta\varepsilon_r)\exp(\beta\varepsilon_\nu)\exp(\beta\varepsilon_e)$$

and, because the various contributions to ε_i in equation (6.21) are independent, then

$$Z = \left[\sum \exp(\beta\varepsilon_t)\right]\left[\sum \exp(\beta\varepsilon_r)\right]\left[\sum \exp(\beta\varepsilon_\nu)\right]\left[\sum \exp(\beta\varepsilon_e)\right] \tag{6.22}$$
$$= Z_t Z_r Z_\nu Z_e.$$

Taking logs we have

$$\ln Z = \ln Z_t + \ln Z_r + \ln Z_\nu + \ln Z_e \tag{6.23}$$

showing that $\ln Z$ consists of a number of additive terms.

6.9 THE FREE ENERGY, F

We have already met the result

$$F = -Nk_{\mathrm{B}}T\ln Z + k_{\mathrm{B}}T\ln N!$$

valid for any MB gas of N molecules (see equation (6.12)).

Using this result together with equation (6.23) we get, for our diatomic gas,

$$F = -Nk_{\mathrm{B}}T\ln Z_t + k_{\mathrm{B}}T\ln N! - Nk_{\mathrm{B}}T\ln Z_r - Nk_{\mathrm{B}}T\ln Z_\nu - Nk_{\mathrm{B}}T\ln Z_e$$
$$= F_t + F_r + F_\nu + F_e.$$

We have written the translational contribution as

$$F_t = -Nk_{\mathrm{B}}T\ln Z_t + k_{\mathrm{B}}T\ln N!$$

which is the same as the total free energy of a monatomic gas.

So for our diatomic gas the total free energy is F_t *plus* the contributions F_r, F_ν and F_e from the internal degrees of freedom.

6.10 THE VARIOUS CONTRIBUTIONS TO Z AND THE OTHER THERMODYNAMIC FUNCTIONS

(a) Z_t (*translational*)
We have

$$Z_t = V \left(\frac{2\pi m k_B T}{h^2} \right)^{3/2}$$

for the translational motion of our diatomic molecule of total mass m. This is the same result as that for a monatomic molecule (see equation (6.6)). The previous results for a monatomic gas

$$U_t = \frac{3}{2} N k_B T$$

and

$$C_{Vt} = \frac{3}{2} N k_B$$

for the translational contribution to the internal energy and heat capacity also apply.

(b) Z_r (*rotational*)
The rotation of our diatomic molecule takes place about an axis through its centre of mass and normal to the line joining the two atoms. Let I be the moment of inertia of the molecule about this axis.

According to quantum mechanics the allowed rotational energy levels are

$$(\varepsilon_r)_j = j(j+1)h^2/8\pi^2 I \tag{6.24}$$

where the quantum number j can have the values $0, 1, 2, \ldots$. For a given value of j there is a degeneracy $g_j = (2j+1)$.

So the rotational partition function is

$$Z_r = \sum_{j=0}^{\infty} (2j+1) \exp[-j(j+1)\theta_r/T] \tag{6.25}$$

(cf. equation (6.5)) where $\theta_r = h^2/8\pi^2 I k_B$ is known as the rotational characteristic temperature.

We first consider the high temperature case $T \gg \theta_r$. Then the energy separation between the quantized levels given by equation (6.24) is $\sim k_B\theta_r \ll k_B T$. Therefore, since $\theta_r/T \ll 1$, the r.h.s. of (6.25) can be replaced by an integral. To proceed we let $j(j+1) = z$; then $(2j+1)dj = dz$. So

$$Z_r = \int_0^\infty \exp(-z\theta_r/T)dz = T/\theta_r. \tag{6.26}$$

Equation (6.26) must be modified slightly if it is to describe homonuclear molecules such as H_2, O_2, N_2 etc. This is because the nuclei of the two atoms are indistinguishable. So, for $T \gg \theta_r$ we can allow for this by writing (6.26) as

$$Z_r = T/\sigma\theta_r \tag{6.27}$$

where the symmetry number σ is equal to 2 for a homonuclear diatomic molecule; for a heteronuclear molecule this correction is not necessary and $\sigma = 1$.

At low temperatures, when $T \ll \theta_r$, the higher rotational states are not excited. This is because the thermal energy $k_B T$ is too small to provide the energy $\sim k_B\theta_r$ needed to go from the ground rotational state $j = 0$ to the higher state $j = 1$. So at $T \ll \theta_r$ the rotational contribution to the heat capacity ~ 0.

The values of θ_r for some gases are given below.

Table 6.1.

	H_2	O_2	N_2	CO	HCl	NO
$\theta_r(K)$	85.4	2.1	2.9	2.8	15.2	2.4

It is found that hydrogen ($\theta_r = 85.4\,K$) and a few other gases can remain in the gaseous state at temperatures below θ_r. All other gases with much lower values of θ_r, such as O_2, N_2 and CO with $\theta_r \sim 2\,K$, will be liquefied at $T = \theta_r$; for such gases to be in the gaseous state T will necessarily be much greater than θ_r and so the high-temperature approximation leading to (6.26) is applicable.

To sum up, for $T \gg \theta_r$,

$$U_r \sim Nk_B T^2\frac{\partial \ln Z_r}{\partial T} \sim Nk_B T$$

and
$$C_{Vr} \sim Nk_B.$$

For $T \ll \theta_r$, $C_{Vr} \sim 0$.

(c) Z_ν (**vibrational**)
A diatomic molecule can be regarded as a one-dimensional harmonic oscillator, the vibrational motion of the two atoms being along the line joining them. The partition function for this vibrational simple harmonic motion is derived in section 9.2, equation (9.1) and is given by

$$Z_\nu = \exp(-\theta_\nu/2T)/[1 - \exp(-\theta_\nu/T)] \tag{6.28}$$

where $\theta_\nu = h\nu/k_B$ is the vibrational characteristic temperature and ν the classical frequency. The energy level spacing of the oscillator is $h\nu$ which is also equal to $k_B\theta_\nu$.

The values of θ_ν for some gases are given below.

Table 6.2.

	H_2	O_2	N_2	CO	HCl	NO
$\theta_\nu(K)$	6140	2239	3352	3080	4150	2690

So, at room temperature, where $T << \theta_\nu$ and $k_B T << k_B \theta_\nu$, the higher vibrational states are not significantly excited and so do not contribute to any extent to the heat capacity. At much higher temperatures, when $T >> \theta_\nu$, the vibrational heat capacity contribution rises to the classical value Nk_B.

To sum up, for $T >> \theta_\nu$, the vibrational energy of the gas $U_\nu \sim Nk_BT$ and the vibrational heat capacity $C_{V\nu} \sim Nk_B$.

For $T << \theta_\nu$, $C_{V\nu} \sim 0$.

(d) Z_e (*electronic*)

The electronic partition function is

$$Z_e = g_0 + g_1 \exp(-\varepsilon'/k_B T) + \ldots \qquad (6.29)$$

(cf. equation (6.5)) where g_0, g_1 etc. are respectively the degeneracies of the ground state, the first excited state, etc. and the energy ε_0 of the ground state has been chosen as the zero of energy. The energy separation of the ground and first excited states is then ε'.

The characteristic temperature for electronic excitation is taken as $\theta_e = \varepsilon'/k_B$. This temperature θ_e is very large compared with room temperature which means that the higher electronic states are not significantly excited at room temperature. For example $\theta_e \sim 120,000K$ for hydrogen.

In general it is true to say that the condition $T << \theta_e$ holds at room temperatures and so, at room temperatures, the only state we need include in (6.29) is the ground state, so that

$$Z_e = g_0.$$

It follows that the higher electronic states do not contribute to any extent to the heat capacity at room temperatures. There are, however, two exceptions to the above general behaviour; these are O_2 and NO. In both of these the two lowest electronic levels are so close together that thermal excitation from the ground state to the first excited state occurs under fairly ordinary conditions. This excitation of the higher state gives an unusual heat capacity contribution (the Schottky anomaly); see section 11.2.

6.11 THE TOTAL PARTITION FUNCTION AND EQUATION OF STATE FOR A DIATOMIC GAS

We return to equation (6.22) which gives the total partition function for one diatomic molecule as

$$Z = Z_t Z_r Z_\nu Z_e.$$

Z_t depends on V and T whereas Z_r, Z_ν and Z_e depend on T alone. We can write

$$Z = Z_t Z_{\text{int}}$$

where $Z_{\text{int}} = Z_r Z_\nu Z_e$ is the contribution from the internal degrees of freedom. So

$$Z = V \left(\frac{2\pi m k_B T}{h^2} \right)^{3/2} Z_{\text{int}}.$$

From equation (6.15a) the pressure P is

$$P = N k_B T \left(\frac{\partial \ln Z}{\partial V} \right)_T$$
$$= \frac{N k_B T}{V}$$

and so we have

$$PV = N k_B T$$

as the equation of state of an ideal diatomic gas.

6.12 THE HEAT CAPACITY OF A DIATOMIC GAS

A rough sketch showing how the heat capacity of a diatomic gas of N molecules varies with temperature is shown in Fig. 6.4; the temperature scale is not linear.

At temperatures just above the boiling point where $T \ll \theta_r$, the only molecular motion is translational and $C_V = \frac{3}{2} N k_B$. As the temperature is raised and approaches θ_r the higher rotational states become excited and when $T \gg \theta_r$ the rotational contribution to the heat capacity is $N k_B$. So in the temperature range $\theta_r \ll T \ll \theta_\nu$ the total heat capacity is

$$C_V = \frac{3}{2} N k_B + N k_B = \frac{5}{2} N k_B$$

and the graph of Fig. 6.4 steps up to this higher value of $\frac{5}{2} N k_B$. It is found that room temperature lies between θ_r and θ_ν so that $C_V \simeq \frac{5}{2} N k_B$, at room temperature, for a diatomic gas.

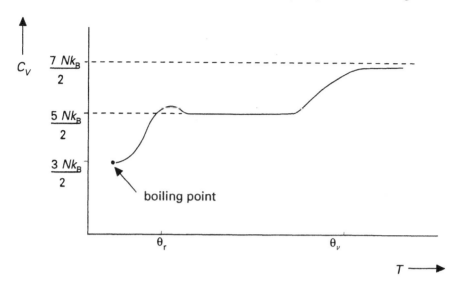

Fig. 6.4. Variation of the heat capacity C_V with temperature for diatomic
gas. Note that the temperature scale is not linear. For H_2, $\theta_r \sim 85\,K$ and
$$\theta_\nu \sim 6000\,K.$$

As T approaches θ_ν, the higher vibrational states begin to be excited and when
$T \gg \theta_\nu$ the vibrational contribution to the heat capacity is Nk_B. So for $T \gg \theta_\nu$,
the total heat capacity is

$$C_V = \frac{3}{2}Nk_B + Nk_B + Nk_B = \frac{7}{2}Nk_B$$

and the graph rises up another step to a value of $\frac{7}{2}Nk_B$.

It must be emphasized that the above is only an outline theory for calculating
C_V for an ideal diatomic gas. For a more detailed treatment of the behaviour of a
particular gas, such as hydrogen, the reader must turn to more advanced texts.

7

Fermi–Dirac gases

7.1 INTRODUCTION

We consider an ideal Fermi–Dirac gas consisting of a total of N fermions each of mass m in a volume V at temperature T. It will be assumed that there are no interactions between the fermions which are particles of half integral spin $\left(\frac{1}{2}, \frac{3}{2}, \dots\right)$.

From equation (5.11) the Fermi–Dirac (FD) distribution may be written as

$$f_k = \frac{N_k}{g_k} = \frac{1}{\exp[(\varepsilon_k - \mu)/k_\mathrm{B}T] + 1} \tag{7.1}$$

where $\mu = \alpha k_\mathrm{B} T$.

In this FD case the chemical potential μ can be identified physically at low temperatures, as we shall see in section 7.2; it is also referred to as the *Fermi energy*. The term on the r.h.s. of equation (7.1) is known as the *Fermi function* which we can write in a more convenient form by treating the energy levels as a continuum, so that we have for the Fermi function

$$f(\varepsilon) = \frac{1}{\exp[(\varepsilon - \mu)/k_\mathrm{B}T] + 1}. \tag{7.2}$$

From equations (7.1) and (7.2) we see that the Fermi function gives the probability that a state with energy ε will be occupied by a fermion. At any temperature $f(\varepsilon)$ has the value $\frac{1}{2}$ for $\varepsilon = \mu$.

If the number of energy states between ε and $\varepsilon + \mathrm{d}\varepsilon$ is $g(\varepsilon)\mathrm{d}\varepsilon$ and the number of fermions with energies in the same energy range is $N(\varepsilon)\mathrm{d}\varepsilon$, then the Fermi function is

$$f(\varepsilon) = \frac{N(\varepsilon)\mathrm{d}\varepsilon}{g(\varepsilon)\mathrm{d}\varepsilon} \qquad \text{cf. equation (7.1)}$$

or

$$N(\varepsilon)\mathrm{d}\varepsilon = f(\varepsilon)g(\varepsilon)\mathrm{d}\varepsilon. \tag{7.3}$$

7.2 THE FERMI ENERGY

We now discuss the significance of the Fermi energy μ. To do this we consider the FD assembly to be at the absolute zero of temperature with the Fermi energy then being μ_0.

The quantity $(\varepsilon - \mu_0)/k_B T$ has two possible values when $T = 0$. When $\varepsilon > \mu_0$, $(\varepsilon - \mu_0)/k_B T = \infty$ and when $\varepsilon < \mu_0$, $(\varepsilon - \mu_0)/k_B T = -\infty$. Thus the two corresponding values of the Fermi function are

(a) $f(\varepsilon) = \frac{1}{e^{\infty} + 1} = 0$ when $\varepsilon > \mu_0$;

(b) $f(\varepsilon) = \frac{1}{e^{-\infty} + 1} = 1$ when $\varepsilon < \mu_0$.

These last two equations tell us that at $T = 0$ the probability that a state with energy $\varepsilon > \mu_0$ is occupied is zero, that is, all states with energy greater than μ_0 will be empty. Conversely the probability that a state with energy $\varepsilon < \mu_0$ is occupied at $T = 0$ is unity, so that all such states are occupied. The variation of $f(\varepsilon)$ with ε at $T = 0$ is shown in Fig. 7.1.

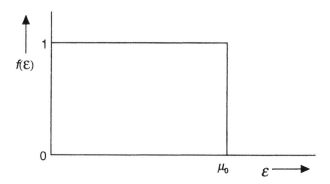

Fig. 7.1. The Fermi function at $T = 0$.

The behaviour of $f(\varepsilon)$ may be explained as follows. At $T = 0$ the fermions will of necessity occupy the lowest energy states which are available to them. Because of the Pauli exclusion principle only one fermion is allowed per state and so all N particles of our assembly will be 'packed into' the N lowest energy states. The

Fermi energy level μ_0 is the energy of the highest occupied state; above this energy level μ_0 the states are unoccupied. We have thus physically identified μ_0.

This means that there is only one distinguishable configuration (that is, microstate) of the whole assembly at $T = 0$. So $\Omega = 1$ and $S = k_B \ln \Omega = 0$ in agreement with the third law of thermodynamics (see, for example, Rosser, Chapter 1).

To find μ_0 we can use the condition $\sum_k N_k = N$ or $\int_0^\infty N(\varepsilon)d\varepsilon = N$ which, from (7.3), gives us

$$\int_0^\infty f(\varepsilon)g(\varepsilon)d\varepsilon = N.$$

Now at $T = 0$, $f(\varepsilon) = 1$ for $\varepsilon \leqslant \mu_0$ and $f(\varepsilon) = 0$ for $\varepsilon > \mu_0$; thus the last equation may be written as

$$\int_0^{\mu_0} g(\varepsilon)d\varepsilon = N. \tag{7.4}$$

For $g(\varepsilon)d\varepsilon$ we use the result

$$g(\varepsilon)d\varepsilon = GV2\pi \left(\frac{2m}{h^2}\right)^{3/2} \varepsilon^{1/2}d\varepsilon \tag{7.5}$$

(see, for example, Rosser, Chapter 12) instead of our previous result in equation (4.6). In equation (7.5), $G = (2s + 1)$ is the spin degeneracy, where s is the spin quantum number; so, for fermions of spin $\frac{1}{2}$, such as electrons, G is equal to 2 and equation (7.5) becomes

$$g(\varepsilon)d\varepsilon = \frac{4\pi V}{h^3}(2m)^{3/2}\varepsilon^{1/2}d\varepsilon. \tag{7.6}$$

For spin $\frac{1}{2}$ fermions there are two spin states $(+\frac{1}{2}$ and $-\frac{1}{2})$ per energy level and so we see why the r.h.s. of our previous equation (4.6) must be multiplied by 2 to give the result in equation (7.6).

From equations (7.4) and (7.6) we get

$$\frac{4\pi V}{h^3}(2m)^{3/2} \int_0^{\mu_0} \varepsilon^{1/2}d\varepsilon = N$$

and so the Fermi energy at $T = 0$ is

$$\mu_0 = \frac{h^2}{2m}\left(\frac{3N}{8\pi V}\right)^{2/3}. \tag{7.7}$$

A convenient way of comparing the Fermi energy with the usual thermal energy $k_B T$ is to write

$$\mu_0 = k_B T_F$$

which defines the *Fermi temperature*, T_F.

7.3 THE ELECTRON GAS IN A METAL

For a full discussion of the free electron model of a metal the reader must turn to books dealing with the solid state, for example, Rosenberg (1989). According to this model the conduction electrons in a metal are able to move freely within the boundaries of the specimen but their escape from the metal is opposed by a potential jump at the boundaries. Effectively they move in a potential box whose walls coincide with these boundaries (see Fig. 7.6). These conduction electrons are pictured as forming a 'gas' of free electrons which behaves as an ideal FD gas.

We now ask: what are typical values for the Fermi energy μ_0 and Fermi temperature T_F for metals? The values of μ_0 can be calculated from equation (7.7) and turn out to be of the order of $5\,\mathrm{eV}$ or $10^{-18}\,\mathrm{J}$. In Table 7.1 estimates of the Fermi temperature $T_F = \mu_0/k_B$ are given for some of the monovalent metals, assuming one free electron per atom.

Table 7.1. Fermi temperatures of some of the monovalent metals

Metal	$T_F \times 10^{-4}(\mathrm{K})$
Li	5.5
Na	3.7
K	2.4
Cu	8.2
Ag	6.4
Au	6.4

We see that T_F is of the order of $10^5\,\mathrm{K}$ and any metal will have melted and vaporized long before its temperature T has approached this value; that is, long before $k_B T$ has reached the value of the Fermi energy.

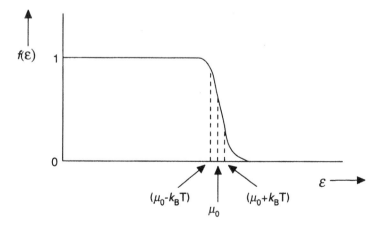

Fig. 7.2. Fermi function for a temperature T where $T_F \gg T > 0$.

We return to consider the graph of $f(\varepsilon)$ against ε. Its shape for $T = 0$ was shown in Fig. 7.1. As T increases, the graph assumes a more rounded shape with only those electrons with energies near μ_0 being affected (Fig. 7.2). States within about $k_B T$ below μ_0 will be partly emptied and those within about $k_B T$ above μ_0 will be partly occupied.

The value μ of the Fermi energy at $T \neq 0$ depends on the temperature; for $k_B T << \mu_0$ (i.e. $T << T_F$) it can be shown that μ is given by

$$\mu \simeq \mu_0 \left[1 - \frac{\pi^2}{12} \left(\frac{k_B T}{\mu_0} \right)^2 \right], \tag{7.8}$$

(see, for example, Guenault, Chapter 8).

So, for $k_B T << \mu_0$, μ is very close to its value μ_0 at $T = 0$ since the second term in the large square bracket is small and we can write either μ or μ_0 in its denominator. When $k_B T << \mu_0$ (i.e. $T << T_F$) the electron gas is *degenerate* and when $k_B T >> \mu_0$ (or $T >> T_F$) it is non-degenerate (the classical limit). For a typical metal at room temperature T the electron gas is therefore degenerate.

Let us now consider equation (7.3), namely

$$N(\varepsilon)d\varepsilon = f(\varepsilon)g(\varepsilon)d\varepsilon.$$

From equation (7.6) we can write $g(\varepsilon) = C\varepsilon^{1/2}$, which tells us that the graph of $g(\varepsilon)$ against ε is a parabola. So the graph of $N(\varepsilon)$ against ε is obtained by 'multiplying' the $(f(\varepsilon), \varepsilon)$ and $(g(\varepsilon), \varepsilon)$ graphs. The result is shown in Fig. 7.3.

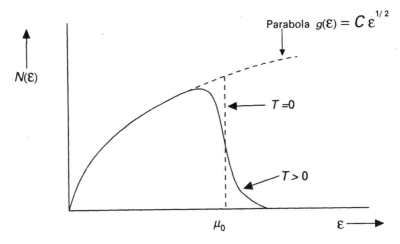

Fig. 7.3. The distribution of electron energies for $T = 0$ and $T_F >> T > 0$.

7.4 THE ELECTRONIC HEAT CAPACITY

At first, at the beginning of this century, the electron gas was assumed to consist of classical particles, each possessing a kinetic energy $\frac{3}{2}k_BT$. This implies that each conduction electron in metals should make a contribution of $\frac{3}{2}k_B$ to the heat capacity, that is, a contribution of $\frac{3}{2}R$ per mole *in addition to* the *lattice* heat capacity of $3R$ per mole (see section 9.1). This, however, is not observed and in fact the electronic contribution at room temperature is found to be only of the order of $10^{-1}R$ to $10^{-2}R$ per mole which is very small compared with the lattice heat capacity. The explanation for such a small value for the electronic heat capacity is one of the great successes of quantum mechanics. We now proceed to discuss this matter.

The internal energy U is the total energy of all the free electrons; so $U = \sum_k N_k \varepsilon_k$

which can be written as

$$U = \int_0^\infty \varepsilon N(\varepsilon) \mathrm{d}\varepsilon$$

$$(7.9)$$

$$= \int_0^\infty \varepsilon f(\varepsilon) g(\varepsilon) \mathrm{d}\varepsilon$$

using equation (7.3).

We now consider, in turn, the cases $T = 0$ and $T_F \gg T > 0$.

(a) $T = 0$

At $T = 0$ equation (7.9) takes the form

$$U_0 = \int_0^{\mu_0} \varepsilon g(\varepsilon) \mathrm{d}\varepsilon$$

since $f(\varepsilon) = 1$ for $\varepsilon \leqslant \mu_0$ and $f(\varepsilon) = 0$ otherwise.

Putting $g(\varepsilon) = C\varepsilon^{1/2}$ (see equation (7.6)) we get

$$U_0 = \int_0^{\mu_0} C\varepsilon^{3/2} \mathrm{d}\varepsilon = \frac{2}{5} C\mu_0^{5/2}.$$

Also, from equation (7.4)

$$N = \int_0^{\mu_0} g(\varepsilon) \mathrm{d}\varepsilon$$

$$= \int_0^{\mu_0} C\varepsilon^{1/2} \mathrm{d}\varepsilon \qquad (7.10)$$

$$= \frac{2}{3} C\mu_0^{3/2}.$$

From the above equations we get

$$U_0 = \frac{3}{5} N \mu_0. \qquad (7.11)$$

U_0 represents the zero point energy and shows us that the mean energy per particle at $T = 0$ is $0.6\mu_0$. U_0 has a large value because the Pauli exclusion principle compels the particles to occupy all energy states ranging from the lowest value up to the Fermi energy level μ_0. This point was discussed previously in section 7.2.

(b) $T_F \gg T > 0$

As the temperature is raised from $T = 0$ to $T = T$ it is assumed that the only electrons elevated in energy are those within about $k_B T$ of μ_0. This number is of the order $g(\mu_0)k_B T$. (We can see this from equation (7.3) in which we substitute $f(\varepsilon) = 1$, $g(\varepsilon) = g(\mu_0)$ and $d\varepsilon = k_B T$). For a temperature T the integral in equation (7.9) can be evaluated using the value of μ in equation (7.8). Details of this evaluation are given for example in Guenault's *Statistical physics*, Chapter 8 and his Appendix 3. The result is

$$U = U_0 + \frac{\pi^2}{6}(k_B T)^2 g(\mu_0)$$

$$= U_0 + U'.$$

The thermal term U' is small compared with U_0. The electronic heat capacity C_e follows at once, since

$$C_e = \frac{dU}{dT}$$

$$= \frac{dU'}{dT}$$

$$= \frac{\pi^2}{3} k_B^2 T g(\mu_0).$$

Now $g(\mu_0) = C\mu_0^{1/2} = 3N/2\mu_0$, using equation (7.10). Also, since $\mu_0 = k_B T_F$, we have

$$C_e = \frac{\pi^2}{2} k_B N \frac{T}{T_F} \qquad (a)$$

$$= \frac{\pi^2}{2} k_B^2 N \frac{T}{\mu_0} \qquad (b) \qquad \right\} \qquad (7.12)$$

$$= \gamma T, \text{ say.} \qquad (c)$$

Equation (7.12(c)) is the form in which the electronic heat capacity is often written. So C_e varies linearly with T in this temperature range.

To illustrate matters we do a calculation for a metal at $T = 300$ K $(<< T_F)$. We take $\mu_0 = 5\,\text{eV} = 8 \times 10^{-19}\,\text{J}$ and substitute into equation (7.12(b)) whence we get $C_e = 0.21\,\text{J K}^{-1}\text{mol}^{-1}$ which is less than 1 % of the lattice heat capacity $3R = 24.9\,\text{J K}^{-1}\text{mol}^{-1}$ (see section 9.1). However, at much lower temperatures, below about 10 K, the picture becomes different. At such temperatures the lattice heat capacity has the form BT^3 and so the total heat capacity is

$$C_V \quad = \quad \underset{\text{(electronic)}}{\gamma T} \quad + \quad \underset{\text{(lattice)}}{BT^3} .$$

At these temperatures the linear term γT is dominant (see Fig. 7.4). So if we wish to determine γ experimentally we must do so at these very low temperatures.

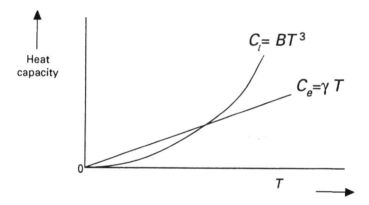

Fig. 7.4. The heat capacity of a metal at low temperatures (C_e is the electronic heat capacity and C_l the lattice heat capacity).

The behaviour of our FD electron gas at sufficiently high temperatures, when T is comparable with the Fermi temperature T_F, becomes classical and the molar electronic heat capacity is of the order $3R/2$ as discussed at the beginning of this section (see also Fig. 7.5(b)).

As we have seen, the electronic heat capacity of a metal at room temperature is very small compared with the lattice heat capacity $3R$. So the heat capacity of a metal is then virtually the same as that of a non-metal.

For further details the reader is referred to Rosenberg's book *The solid state* (1989), Third Edition and Kittel's book *Introduction to solid state physics* (1971), Fourth Edition.

7.5 THE THERMODYNAMIC FUNCTIONS OF AN IDEAL FERMION GAS

Although we have been concentrating on a gas of free electrons in the previous two sections, most of the results which we met there are valid for an ideal fermion gas.

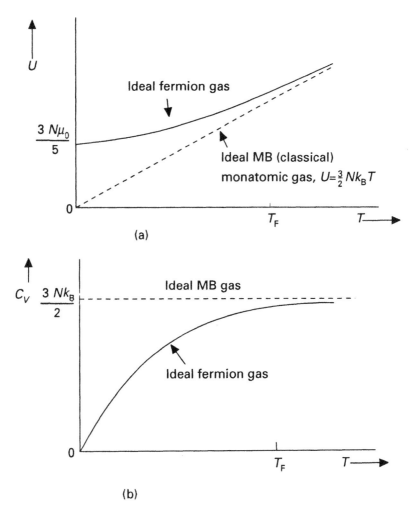

Fig. 7.5 (a) the (U, T) curve and (b) the (C_V, T) curve for an ideal FD gas. Since $P = 2U/3V$, that is $P \propto U$, the (P, T) curve is similar to the (U, T) curve; it starts at $P_0 = 2N\mu_0/5V$ at $T = 0$.

Take first the internal energy U of the fermion gas. Its variation with absolute temperature T at constant volume V is shown as the full-line curve in Fig. 7.5(a). The curve starts from the value $\frac{3}{5}N\mu_0$ at $T = 0$ (see equation (7.11)). When T

exceeds the Fermi temperature T_F the behaviour of the fermion gas approximates more and more to that of an ideal MB gas whose internal energy is represented by the dotted line $U = \frac{3}{2}Nk_BT$. We see that the value of U for our ideal fermion gas is always greater than that of an ideal MB gas for which the values of V, T and N are the same. The fermion gas is degenerate in the temperature range $T < T_F$ where its behaviour departs from that of the ideal MB gas.

Next consider the heat capacity at constant volume given by $C_V = \mathrm{d}U/\mathrm{d}T$. Its variation with temperature is shown in Fig. 7.5(b). For low values of temperature when $T << T_F$ we have seen that the value of the heat capacity is the value of C_e in equations (7.12) and for such temperatures the heat capacity varies linearly with T. When $T > T_F$, the heat capacity approaches the classical value of $\frac{3}{2}Nk_B$ for an ideal MB gas.

Finally we consider the pressure P. It is shown in section 11.1 that $P = 2U/3V$ so that $P \propto U$ if V is a constant. Hence the variation of P with T is the same as that of U with T. Again we see that the pressure of an ideal fermion gas is always greater than that of an ideal MB gas having the same values of V, T and N. There is a zero point pressure of

$$P_0 = \frac{2U_0}{3V}$$

$$= \frac{2N\mu_0}{5V} \tag{7.13}$$

(using equation (7.11)).

This pressure can have extremely high values as we shall see in the next section.

7.6 SOME SIMPLE CALCULATIONS

To obtain an estimate of some of the quantities involved we now carry out some calculations for silver.

From Table 7.1 we have, for silver,

$$T_F = 6.4 \times 10^4 \text{ K}.$$

Therefore $\mu_0 = k_B T_F \simeq 9 \times 10^{-19}$ J$\simeq 5.6$ eV since 1 eV$= 1.6 \times 10^{-19}$ J.

Hence the mean energy per electron at $T = 0$ is, from equation (7.11), given by

$$\bar{\varepsilon} = \frac{U_0}{N} = \frac{3}{5}\mu_0 \simeq 5.4 \times 10^{-19} \text{ J} = 3.4 \text{ eV}.$$

This compares with the value of only 0.04 eV for the mean kinetic energy $\frac{3}{2}k_BT$ of a gas molecule at 300 K.

The pressure of the electron gas at $T = 0$ is, from equation (7.13), given by

$$P_0 = \frac{2N\mu_0}{5V}.$$

For silver $N/V \simeq 6 \times 10^{28}$ electrons m^{-3} and so

$$P_0 \simeq 22 \times 10^9 \, \text{Nm}^{-2}$$
$$\simeq 2.2 \times 10^5 \, \text{atm.}$$

As we see from the graph of Fig. 7.5(a) the pressure of the electron gas at *room temperature* would be even greater than P_0. Although this electron gas pressure is so very high the electrons do not 'evaporate' from the metal since they are prevented from doing so by the potential barrier at the surface (see Fig. 7.6 and also section 7.3).

A simple diagram of the potential box used in the free electron model of metals is given in Fig. 7.6. The energy required to remove an electron to the outside of the metal is equal to ϕ which is known as the work function and is of the order of 3 eV. So the total depth of the potential box in a metal is about 8 eV since the Fermi energy level is about 5 eV above the bottom of the box.

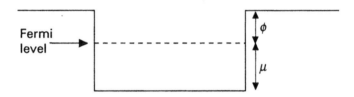

Fig. 7.6. The potential box used in the free electron theory of metals.

We return to consider the very high pressure of the electron gas. Such a high pressure also exists in the degenerate electron gas in a white dwarf star; in this case the gas has a pressure of the order of 10^{17} atm and this huge pressure prevents the gravitational collapse of the core of the star.

For further reading see Rosser's *Statistical physics*, (1982) Chapter 12.

7.7 PAULI PARAMAGNETISM

A material placed in a magnetic field acquires an induced magnetic moment whose direction is either parallel to the applied field or is such that it opposes the field. The former case is that of paramagnetism (and ferromagnetism) and the latter that of diamagnetism. While diamagnetism occurs in all materials, weak paramagnetism occurs in all metals. This paramagnetic effect is temperature-independent. In this section we shall consider how to explain this effect using the free electron theory discussed in section sections 7.3 and 7.4.

The magnetic moments of the individual atoms are due to the spins of the free electrons. In the absence of a magnetic field these elementary dipoles will exist in random directions but when a magnetic field B is applied to the metal these dipoles will line up parallel or antiparallel to the field.

If the magnetic moment due to the spin of the electron is μ_B (the Bohr magneton), the magnetic energy of the electron will be $-\mu_B B$ when it is parallel to B and $\mu_B B$ when it is antiparallel to B. So a parallel dipole will have an energy which is less, by $2\mu_B B$, than that of an antiparallel dipole. Because of this energy difference some of the electrons will reorientate themselves until there are more parallel dipoles than antiparallel, thus reducing the overall energy of the arrangement. There will therefore be a resultant magnetic moment parallel to B giving the paramagnetic effect.

Now consider the energy level diagram at $T = 0\,\mathrm{K}$ for the two states of the electron spin in the absence of an applied field. It is shown as a parabolic 'box' in Fig. 7.7, the occupied states being shown shaded. Each half of the graph is similar to the parabola $g(\varepsilon) = C\varepsilon^{1/2}$ shown in Fig. 7.3 except that the energy ε is now plotted as ordinate. The density of states $g(\varepsilon)$ is that given by equation (7.6). In the graph the number of states is plotted as $\frac{1}{2}g(\varepsilon)$, since only a half of the electrons will be in one spin state (spin $+\frac{1}{2}$) and a half in the other (spin $-\frac{1}{2}$).

Each half of the parabolic box is filled up to ε_F, the Fermi level at $T = 0\,\mathrm{K}$. (We use ε_F here for the Fermi level rather than μ_0 as previously; this is to avoid confusion by using too many μs in this section.)

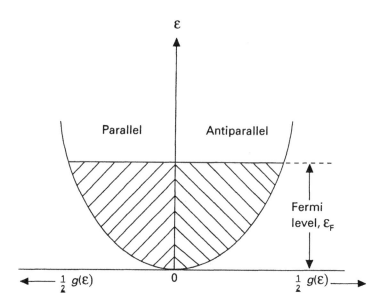

Fig. 7.7. The energy levels in zero magnetic field at $T = 0$.

Next, suppose that a magnetic field B is applied. This will result, initially, in a shift in the two halves of the graph of Fig. 7.7; this displacement is shown in Fig. 7.8. The parallel moments will take up energies lower by an amount $\mu_B B$ while the antiparallel ones have their energies increased by the same amount. The two halves of the parabola will thus be displaced by a vertical distance $O_1 O_2$ where $O_1 O_2 = 2\mu_B B$.

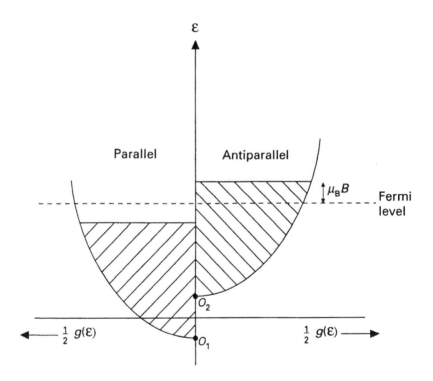

Fig. 7.8. Initial arrangement of energy states after a field B is applied.

For equilibrium the electrons in the energy band of thickness $\mu_B B$ at the top of the 'antiparallel' side of the graph of Fig. 7.8 must reverse their orientations, thus reducing their energies, to occupy the top of the 'parallel' side of the graph. This final state of affairs is shown in Fig. 7.9. The number of electrons moving across in this reorientation is given by

$$\Delta N = f(\varepsilon) \times \frac{1}{2} g(\varepsilon) \mathrm{d}\varepsilon$$

(cf. equation (7.3)). In this case $f(\varepsilon) = 1$, $\varepsilon = \varepsilon_F$ and $\mathrm{d}\varepsilon = \mu_B B$, so

$$\Delta N = \frac{1}{2} g(\varepsilon_F) \mu_B B.$$

In this reorientation each of the ΔN electrons changes its moment from $+\mu_\mathrm{B}$ to $-\mu_\mathrm{B}$, a total change of $2\mu_\mathrm{B}$. So the net magnetic moment of the specimen will be

$$2\mu_\mathrm{B}\Delta N = \mu_\mathrm{B}^2 g(\varepsilon_\mathrm{F})B$$

and this moment is in the same direction as B.

The paramagnetic susceptibility χ_P is defined as

$$\chi_\mathrm{P} = \frac{\text{magnetic moment}}{\text{field intensity } H}$$

where the field intensity H is given by $B = \mu_0 H$ and μ_0 is here the permeability of free space. So

$$\chi_\mathrm{P} = \mu_\mathrm{B}^2 g(\varepsilon_\mathrm{F})\mu_0. \tag{7.14}$$

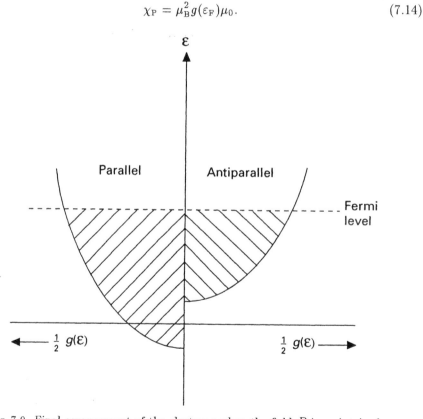

Fig. 7.9. Final arrangement of the electrons when the field B is maintained.

This type of paramagnetism is often called Pauli paramagnetism. It occurs because of the ability of the small number of electrons whose energies are close to ε_F (those in the top of the r.h.s. of Fig. 7.8) to reverse their orientations. Above $T = 0$ any variation in χ_P with temperature depends on the way in which the Fermi energy varies with temperature and this variation is quite negligible. So, in effect, χ_P is temperature independent.

We should finally note that there is also present a *diamagnetic* contribution of $-\frac{1}{3}\chi_P$ to the susceptibility of the electron gas. This reduces the χ_P in equation (7.14) to about two-thirds of its value.

For a fuller discussion the reader is referred to Rosenberg's *The solid state*, Chapter 7 and similar standard books.

7.8 THE LIQUID HELIUM PROBLEM

The two isotopes of helium, ^{4}He and ^{3}He, have atomic mass numbers of 4 and 3 respectively. The ^{4}He atom consists of a nucleus containing four nucleons (two protons and two neutrons) together with two extranuclear electrons; each of these six particles is a fermion, having half-integral spin. So, since the ^{4}He atom contains an even number of fermions, its total spin is integral and it should be described by symmetric wave functions and obey Bose–Einstein statistics.

The ^{3}He atom has three nucleons (two protons and one neutron) in the nucleus and two extranuclear electrons; since each of these particles is a fermion with half-integral spin, the total spin must be half-integral. So the ^{3}He atom is described by antisymmetric wave functions and obeys Fermi–Dirac statistics.

Liquid ^{4}He and ^{3}He will remain liquid under their own vapour pressure down to $0\,$K. Their atoms have 'closed' electronic shells and so are chemically inert; hence the interatomic forces (which are of van der Waals type) are very weak. We picture our liquid helium as an assembly of particles forming a quasi-lattice in which the interatomic spacing is a. From Heisenberg's Uncertainty Principle $\Delta p_x \cdot \Delta x \sim h$. Since $\varepsilon = p^2/2m$ for a non-relativistic particle we see that the uncertainty principle demands a zero-point kinetic energy per atom of $\varepsilon \sim h^2/ma^2$. For a given value of a, ε is greatest for a small m, as is the case for helium. In other words, the zero point energy for helium is comparatively large and is a consequence of the uncertainty principle.

So for liquid helium at these near-zero temperatures we have two effects present. Firstly we have the binding energy due to the interatomic forces; this tends to pull the atoms together. This is opposed by the large zero point energies which produce repulsive forces between the atoms. In this situation the binding and zero point energies are comparable whereas, usually, in other substances the zero point energy is much less than the binding energy. The result is that the liquid helium is 'inflated' to a volume much larger than one would expect. The late Sir Francis Simon put it in a vivid way when he said that 'liquid helium is blown up by its own zero point energy'. So liquid helium has an 'open' rather than a close-packed structure and it is therefore necessary to apply a considerable external pressure to cause it to solidify; this is true of both liquid isotopes. The zero point energy of ^{3}He is greater than that of ^{4}He because of the smaller atomic mass number. So the vapour pressure of liquid ^{3}He is higher than that of liquid ^{4}He which means that a higher pressure is needed to solidify it.

As a result of the above facts, both liquids ^{4}He and ^{3}He have been treated as 'gases', the density of the gases being the same as that of the actual liquids. We shall discuss this point further in the next section.

7.9 LIQUID ^{3}HE AS A FERMION GAS

The ^{3}He atom has a nuclear spin of $\frac{1}{2}$ and so an assembly of such atoms constitutes a fermion gas. Because of this nuclear spin, liquid ^{3}He will exhibit nuclear paramagnetism and the susceptibility χ has been measured down to about 1 K. It shows a Curie type variation $\chi = a/T$. At these low temperatures, in the absence of an applied magnetic field, the orientation of the nuclear spins is quite a random one. This spin disorder gives rise to a fairly large entropy in the liquid.

Consider the Clausius–Clapeyron equation

$$\frac{dP}{dT} = \frac{\Delta S}{\Delta V} \tag{7.15}$$

which relates the slope dP/dT of the melting curve to ΔS and ΔV, the entropy and volume changes on melting. In almost all instances the entropy and volume increase when a substance melts, so that both ΔS and ΔV are positive. However, in the case of ^{3}He the entropy does not behave in this 'normal' manner below 0.3 K. To see this consider the melting curve of ^{3}He shown in Fig. 7.10.

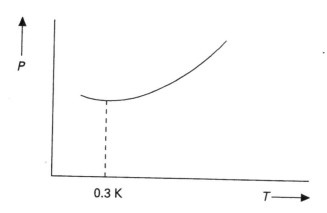

Fig. 7.10. The melting curve of ^{3}He.

It shows that dP/dT becomes negative below 0.3 K. From equation (7.15) this implies that either ΔS or ΔV is negative. However, since we know from experiment that ΔV is positive it follows that ΔS must be negative; in other words the liquid has less entropy, and is therefore less disordered, than the solid as we go along the melting curve below 0.3 K.

We conclude that the nuclear spins in the solid are more disordered, which accounts for the greater entropy. As far as the situation in liquid ^{3}He is concerned the only spins free to change their orientation are those within a small energy band at the Fermi level. (We met a similar situation when discussing the electronic heat capacity (section 7.4) and Pauli paramagnetism (section 7.7).) The fact that only this relatively small number of spins can take part in the disorder process accounts for the entropy of the liquid being lower than that of the solid for temperatures below 0.3 K.

Liquid ^{3}He has a very large molar volume whose atoms are far apart. So it can be considered as a fermion gas of the same density. The heat capacity and entropy of liquid ^{3}He decay to zero in much the same way as they do for this fermion gas.

A more detailed theory for treating ^{3}He liquid as a gas, due to Landau and others, takes account of correlations between the movements of the atoms. This is achieved by replacing the ^{3}He atoms by ^{3}He 'quasi-particles' obeying the FD statistics. The effective mass of a quasi-particle is some three (or more, depending on the density) times that of the ^{3}He atom (see Guenault's *Statistical physics* (1988), Chapter 8).

7.10 APPLICATION TO SEMICONDUCTORS

We start with a brief description of semiconductors, fuller details being available in the various books on solid state physics, e.g. Rosenberg's *The solid state* (1989), Third Edition.

Semiconductors are electronic conductors which at room temperatures have an electrical resistivity in the range 10^{-4} to $10^7\,\Omega\,\mathrm{m}$. These values lie between those for conductors ($\sim 10^{-7}\,\Omega\,\mathrm{cm}$) and those for insulators ($\sim 10^{12} - 10^{20}\,\Omega\,\mathrm{m}$). At 0 K a pure semiconductor acts as an insulator and at this absolute zero it is pictured as having a vacant 'conduction' band separated from a filled 'valence' band by an energy gap ε_g as shown in Fig. 7.11.

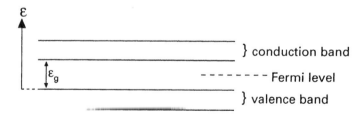

Fig. 7.11. The electron bands in a pure semiconductor. At 0 K all the states in the valence band are filled and all those in the conduction band are vacant and the conductivity is zero. As the temperature rises thermal excitation from the valence to the conduction bands occurs across the energy gap ε_g.

As the temperature rises, some of the electrons in the valence band are thermally excited to the conduction band. These electrons in the conduction band and also the holes they leave in the valence band are mobile and they will both contribute to the electrical conductivity when an external electric field is applied. In our pure semiconductor this is known as *intrinsic* conductivity. The best known semiconducting materials are silicon, germanium and GaAs.

In order to calculate the intrinsic conductivity of a pure semiconductor at a given temperature T we need to find the equilibrium value of N_e, the number of electrons per unit volume in the conduction band and N_h, the equilibrium number per unit volume of holes in the valence band. Of course $N_e = N_h$.

It is shown in books on the physics of solids that the conductivity is given by

$$\sigma = N_e e \mu_e + N_h e \mu_h$$

where e is the value of the electronic charge and μ_e, μ_h are the mobilities for electrons and holes respectively.

We now show how statistical mechanics enables us to calculate N_e and N_h.

First, let us consider N_e. We take the top of the valence band in Fig. 7.11 as the zero of energy; ε_g is then the energy at the bottom of the conduction band. The Fermi level ε_F is, as we shall show, in the middle of the energy gap in Fig. 7.11.

If ε is the energy of an electron in the conduction band we assume that $(\varepsilon - \varepsilon_F) \gg k_B T$ so that the Fermi function

$$f_e(\varepsilon) = \frac{1}{\exp[(\varepsilon - \varepsilon_F)/k_B T] + 1} \tag{7.16}$$

of equation (7.2) reduces to

$$f_e(\varepsilon) = e^{(\varepsilon_F - \varepsilon)/k_B T}.$$

In all this we are assuming that the electrons in the conduction band are behaving as free electrons. To find the number of states with energy between ε and $\varepsilon + d\varepsilon$ we use the result given in equation (7.6), namely,

$$g(\varepsilon)d\varepsilon = 4\pi V \left(\frac{2m}{h^2}\right)^{3/2} \varepsilon^{1/2} d\varepsilon.$$

In the present case we amend this by using $(\varepsilon - \varepsilon_g)$ instead of ε since we measure the electron energy from the bottom of the conduction band; we also put $m = m_e^*$ where m_e^* is the effective mass of the electron in the band and we put $V = 1$. We then have

$$g_e(\varepsilon)d\varepsilon = 4\pi \left(\frac{2m_e^*}{h^2}\right)^{3/2} (\varepsilon - \varepsilon_g)^{1/2} d\varepsilon \tag{7.17}$$

for the number of states per unit volume with energy between ε and $\varepsilon + d\varepsilon$.

From equation (7.3) the number of electrons per unit volume in the conduction band is

$$N_e = \int_{\varepsilon_g}^{\infty} f_e(\varepsilon) g_e(\varepsilon) d\varepsilon$$

$$= 4\pi \left(\frac{2m_e^*}{h^2}\right)^{3/2} e^{\varepsilon_F/k_B T} \int_{\varepsilon_g}^{\infty} (\varepsilon - \varepsilon_g)^{1/2} e^{-\varepsilon/k_B T} d\varepsilon \qquad (7.18)$$

$$= 2 \left(\frac{2\pi m_e^* k_B T}{h^2}\right)^{3/2} e^{(\varepsilon_F - \varepsilon_g)/k_B T}.$$

Next, consider the concentration of holes, N_h, in the valence band. Since a hole is really the absence of an electron, the distribution function f_h for holes is given by

$$f_h = 1 - f_e$$

where f_e is the electron distribution given by (7.16). So

$$f_h = 1 - \frac{1}{e^{(\varepsilon - \varepsilon_F)/k_B T} + 1}$$

$$= \frac{1}{e^{(\varepsilon_F - \varepsilon)/k_B T} + 1}$$

$$\simeq e^{(\varepsilon - \varepsilon_F)/k_B T}$$

if $(\varepsilon_F - \varepsilon) \gg k_B T$.

Assuming that the holes are near the top of the valence band and that they behave as free particles with an effective mass of m_h^* we can write down an equation similar to equation (7.17) for the holes. This is

$$g_h(\varepsilon) d\varepsilon = 4\pi \left(\frac{2m_h^*}{h^2}\right)^{3/2} (-\varepsilon)^{1/2} d\varepsilon$$

since we measure energy positive upwards from the top of the valence band.

So, if we carry out the same procedure as for the electrons we obtain for the number of holes per unit volume in the valence band

$$N_h = \int_{-\infty}^{0} f_h(\varepsilon) g_h(\varepsilon) d\varepsilon$$

$$(7.19)$$

$$- 2 \left(\frac{2\pi m_h^* k_B T}{h^2} \right)^{3/2} e^{-\varepsilon_F / k_B T}.$$

For our intrinsic semiconductor $N_e = N_h$ and we therefore get, from equations (7.18) and (7.19),

$$\varepsilon_F = \frac{1}{2}\varepsilon_g + \frac{3}{4} k_B T \, \ln(m_h^*/m_e^*).$$

If $m_e^* = m_h^* = m^*$, then $\varepsilon_F = \frac{1}{2}\varepsilon_g$; that is, the Fermi level is in the middle of the energy gap in Fig. 7.11.

So

$$N_e = N_h = \text{ const.} \exp(-\varepsilon_g/2k_B T)$$

showing that both N_e and N_h depend on $\exp(-\varepsilon_g/2k_B T)$ and not on $\exp(-\varepsilon_g/k_B T)$ as we might have expected on the face of it.

For Ge and Si, ε_g has the values of 0.75 eV and 1.1 eV respectively. At room temperature we have $N_e = N_h = 10^{19} \, \text{m}^{-3}$ for Ge and $3 \times 10^{16} \, \text{m}^{-3}$ for Si.

7.11 SUMMARY

In this chapter we have considered the behaviour of an ideal fermion gas and compared it with that of an ideal MB gas. The significance of the Fermi energy level and Fermi temperature have been discussed. We then considered the application of these ideas to various problems, namely, the electron gas in a metal, the phenomenon of paramagnetism, the properties of liquid ^{3}He and the conductivity of semiconductors.

8

Bose–Einstein gases

8.1 INTRODUCTION

We now consider an ideal Bose–Einstein gas consisting of a total of N bosons each of mass m in a volume V at temperature T. We assume that there are no interactions between the bosons which are particles of integral spin $(0, 1, 2, \dots)$.

We start by reminding ourselves of the Bose–Einstein distribution in equation (5.12). This tells us that

$$f_k = \frac{N_k}{g_k} = \frac{1}{e^{-\alpha}e^{-\beta\varepsilon_k} - 1} \tag{8.1a}$$

In this equation f_k is the mean number of bosons in a single energy state (level). If we put $e^{\alpha} = A$ then the value of α (or A) is determined from the condition $\sum_k N_k = N$. So we may write

$$f_k = \frac{1}{\frac{1}{A}e^{\varepsilon_k/k_{\mathrm{B}}T} - 1} \tag{8.1b}$$

since $\beta = -1/k_{\mathrm{B}}T$.

We take the ground state to have zero energy $\varepsilon_k = \varepsilon_0 = 0$. To be physically meaningful f_k must be positive for this ground state as indeed for all higher energy states. If we put $\varepsilon_k = 0$ in (8.1b) then f_k is positive if $\frac{1}{A} > 1$. So $\frac{1}{A}$ must be greater than 1 for (8.1b) to describe our boson gas.

When $\frac{1}{A} >> 1$ (or $A << 1$) then $\frac{1}{A}e^{\varepsilon_k/k_{\mathrm{B}}T} >> 1$ in (8.1b) and $f_k = Ae^{\beta\varepsilon_k}$, the MB or classical result (see equation (4.14)).

We have seen previously (equation (6.7)) how A is given in terms of other quantities. If the density N/V is low or the temperature T high then $A \ll 1$, the MB limit applies and the gas is said to be non-degenerate (because the number of energy states far exceeds the number of particles to occupy them). We have previously considered the value of $A = e^{\alpha}$ in section 6.4.

From (8.1a) the number, N_0, of bosons at any temperature in the ground state is

$$N_0 = \frac{1}{\frac{1}{A} - 1} \qquad (8.2)$$

where, since there is only one state at zero energy, $g_0 = 1$.

As $T \to 0$ all the N bosons will tend to crowd into the ground state, that is, $N_0 \to N$. So

$$\lim_{T \to 0} \frac{1}{\frac{1}{A} - 1} = N.$$

If we treat the energy levels as a continuum we can write (8.1a) as

$$f(\varepsilon) = \frac{1}{\frac{1}{A} e^{-\beta \varepsilon} - 1} \qquad (8.1c)$$

with A determined from the condition

$$\int_0^{\infty} N(\varepsilon) d\varepsilon = N \qquad (8.3)$$

where

$$N(\varepsilon) d\varepsilon = f(\varepsilon) g(\varepsilon) d\varepsilon. \qquad (8.4)$$

(cf. equation (7.3)).

Here $N(\varepsilon)d\varepsilon$ is the number of bosons with energies between ε and $\varepsilon + d\varepsilon$, and $g(\varepsilon)d\varepsilon$ is the number of available energy states in this same energy range. From equation (4.6) we have

$$\left.\begin{array}{l} g(\varepsilon)d\varepsilon = \dfrac{2\pi V}{h^3}(2m)^{3/2}\varepsilon^{1/2}d\varepsilon \\[4mm] \qquad = C\varepsilon^{1/2}d\varepsilon \ \text{(say)} \end{array}\right\} \qquad (8.5)$$

if, for simplicity, we consider spin 0 bosons only; for these the spin degeneracy $G = 1$, see equation (7.5).

From (8.1c) and (8.4) we get

$$N(\varepsilon)d\varepsilon = \frac{g(\varepsilon)d\varepsilon}{\frac{1}{A} e^{-\beta \varepsilon} - 1} \qquad (8.5a)$$

and equation (8.3) becomes

$$N = \int_0^\infty \frac{g(\varepsilon)\mathrm{d}\varepsilon}{\frac{1}{A}\mathrm{e}^{-\beta\varepsilon} - 1}. \tag{8.6}$$

8.2 THE BOSE–EINSTEIN CONDENSATION

Now consider what happens as the value of A increases towards unity. In that case the condition $\frac{1}{A}\mathrm{e}^{\varepsilon/k_\mathrm{B}T} \gg 1$ holds only for very large values of ε. So the gas will no longer behave as an MB gas. At $T = 0$ all the bosons will occupy the ground state $\varepsilon = 0$; for slightly higher values of T some of these bosons will occupy the higher or excited energy states. Now for the lowest energy state $\varepsilon = 0$, the value of $g(\varepsilon) = C\varepsilon^{1/2}$ vanishes so that the corresponding value of $g(\varepsilon)\mathrm{d}\varepsilon$ in (8.5) is zero although, in fact, we know that there will be one state at this zero of energy. So the single ground state $\varepsilon = 0$ is not included in the integral in (8.6), which really only gives N_ex, the number of bosons in the excited states.

Hence the total number N of bosons is made up of N_0 in the ground state and N_ex in the excited states; thus

$$N = N_0 + N_\mathrm{ex}$$

$$= N_0 + \int_0^\infty \frac{g(\varepsilon)\mathrm{d}\varepsilon}{\frac{1}{A}\mathrm{e}^{-\beta\varepsilon} - 1}. \tag{8.7}$$

where, from (8.2),

$$N_0 = \frac{1}{\frac{1}{A} - 1} = \frac{A}{1 - A}. \tag{8.2a}$$

As A increases to 1, N_0 increases to include a large fraction of all the bosons. This increase in N_0 is called a Bose–Einstein condensation.

It is found that the equation $N = N_0 + N_\mathrm{ex}$ behaves very differently above and below a certain temperature T_B, known as the Bose temperature. Above T_B, the term N_ex predominates and can contain all the bosons, that is, N_ex can be as large as N. As T falls below T_B more and more bosons crowd into the ground state and at $T = 0$ they all occupy this state, that is, $N_0 = N$. This 'collapse' of the bosons into the ground state at $T = T_\mathrm{B}$ is a sudden event. Just as in the everyday process of condensation there is a reduction in the volume of geometrical space occupied by the molecules, so is the crowding of bosons into the ground state regarded as a condensation or 'ordering' in momentum space.

We can estimate the temperature T_B as follows. The number N_ex of bosons in the excited states is given by the second term on the r.h.s. of (8.7), that is

$$N_\mathrm{ex} = \int_0^\infty \frac{g(\varepsilon)\mathrm{d}\varepsilon}{\frac{1}{A}\mathrm{e}^{-\beta\varepsilon} - 1}. \tag{8.7a}$$

At temperatures close to the absolute zero, if N_0 is to be a large number, then A must be very close to 1 (equation (8.2a)). So it is a good approximation to assume that A is equal to unity in (8.7a). Remembering that $\beta = -1/k_B T$ and that $g(\varepsilon)$ is given by (8.5) we put $x = \varepsilon/k_B T$. With this substitution we get

$$N_{ex} = V \frac{2\pi}{h^3} (2mk_B T)^{3/2} \int_0^\infty \frac{x^{1/2} dx}{e^x - 1}.$$

The value of the definite integral is $2.612\pi^{1/2}/2$. Therefore

$$N_{ex} = 2.612V \left(\frac{2\pi mk_B T}{h^2} \right)^{3/2}. \tag{8.8}$$

At the temperature T_B all the bosons should be in excited states. So putting N_{ex} equal to N and $T = T_B$ in (8.8) we get

$$N = 2.612V \left(\frac{2\pi mk_B T_B}{h^2} \right)^{3/2}. \tag{8.9}$$

Rearranging this gives us

$$T_B = \frac{h^2}{2\pi mk_B} \left(\frac{N}{2.612V} \right)^{2/3} \tag{8.10}$$

for our Bose temperature.

If we divide (8.8) and (8.9) we get

$$\frac{N_{ex}}{N} = \left(\frac{T}{T_B} \right)^{3/2}$$

which means that the number of bosons in the ground state $\varepsilon = 0$ is

$$N_0 = N - N_{ex}$$
$$= N \left(1 - \left[\frac{T}{T_B} \right]^{3/2} \right). \tag{8.11}$$

The graph of N_0 against T is shown in Fig. 8.1. For $T \leqslant T_B$, N_0 is given by equation (8.11) and for $T > T_B$, $N_0 \sim 0$.

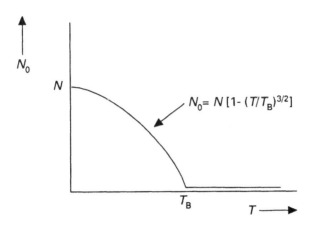

Fig. 8.1. The variation with temperature of the number of bosons, N_0,
in the ground state for an ideal BE gas.

Mathematical note

The argument by which we have replaced $\sum_k N_k = \sum_k f_k g_k$ by an integral like (8.6)
is good enough for many problems in statistical mechanics such as radiation in an
enclosure (section 8.5), the Debye theory of solids (section 9.4) and the Fermi–
Dirac gas (Chapter 7). However, in many mathematical problems it is *not* correct
to replace a series by an integral. Many series can be summed asymptotically as
the corresponding integral plus a set of correcting terms. In the present case it is
sufficient to add in just the term (8.2), all other corrections being negligible for a
large assembly.

This is the mathematical argument which leads us to (8.7) instead of (8.6). The
condensation effect occurs because the integral (8.6) converges even when $A = 1$ so
that we have to introduce the first term of (8.7) in order to get mathematical and
physical sense.

8.3 THE THERMODYNAMIC FUNCTIONS OF AN IDEAL BOSON GAS

The particles in the state $\varepsilon = 0$, at any temperature lower than T_B, make no
contribution to the internal energy, heat capacity, etc.

To make progress we follow the argument of MacDonald (1963). Below T_B,
as we saw in the previous section, the number of bosons in the excited states is

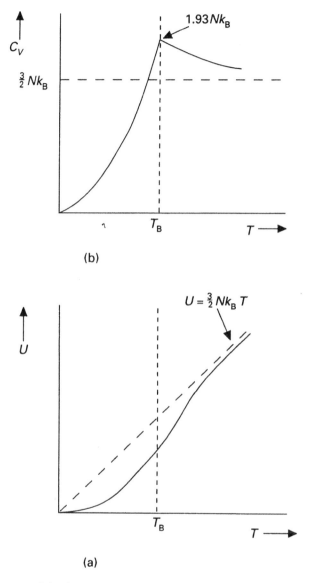

Fig. 8.2. The (a) (U, T) and (b) (C_V, T) curves for an ideal BE gas. Since $P = 2U/3V$, the pressure P varies with temperature in the same way as U.

$N_{\mathrm{ex}} = N(T/T_\mathrm{B})^{3/2}$. As a first approximation we assume that each of these N_{ex} bosons will have an average thermal energy of the order of $k_\mathrm{B}T$. So the internal energy below T_B is about

$$U = Nk_\mathrm{B}T(T/T_\mathrm{B})^{3/2}$$

showing that U varies as $T^{5/2}$ below T_B. The heat capacity $C_V = \mathrm{d}U/\mathrm{d}T$ then varies as $T^{3/2}$ up to a maximum of $1.93Nk_B$ at $T = T_B$. We can see this from the more exact result given by London (1938), namely

$$U = 0.771Nk_B T (T/T_B)^{3/2}$$

for $T \leqslant T_B$. This gives, on differentiating

$$C_V = \frac{\mathrm{d}U}{\mathrm{d}T} = 1.93Nk_B \left(\frac{T}{T_B}\right)^{3/2}$$

for $T \leqslant T_B$; when $T = T_B$, $C_V = 1.93Nk_B$.

When $T \gg T_B$, U tends to its classical value of $3Nk_B T/2$ and C_V decreases from its maximum at $T = T_B$ to its classical value of $3Nk_B/2$. The (U, T) and (C_V, T) graphs are shown in Fig. 8.2. C_V exhibits a cusp maximum at $T = T_B$. As pointed out by MacDonald this cusp is a direct consequence of the onset at $T = T_B$ of the condensation of particles into the ground state.

Since $P = 2U/3V$ (see section 11.1) the pressure P varies with temperature in the same way as U.

8.4 APPLICATION TO ⁴HE

The ^{4}He atom has a resultant spin of zero and is a boson. The phase diagram of ^{4}He is shown in Fig. 8.3. The critical temperature is 5.2 K; whatever the external pressure ^{4}He cannot exist as a liquid above 5.2 K. Liquid ^{4}He remains liquid as T tends to zero at pressures below 25 atm. At a certain temperature there is a large anomaly in the heat capacity which, for liquid ^{4}He under its own saturated vapour pressure is at 2.17 K; this temperature of 2.17 K is known as the λ-point (see Figs 8.3 and 8.4). It is found that there is a coexistence curve in the phase diagram, known as the λ-line, which separates two liquid phases. At temperatures greater than those corresponding to the λ-line the liquid is called liquid He I; its behaviour does not differ greatly from that of other liquids. At temperatures lower than those corresponding to the λ-line the liquid is called liquid He II, and its properties are unique. In particular its viscosity, if measured by its flow rate through a fine capillary, is virtually zero. We say that it is a superfluid or that it possesses superfluidity.

These matters can be described by the so-called two-fluid model of liquid He II. According to this model, liquid He II is regarded as a mixture of two interpenetrating fluids known as the normal and superfluid components. The normal component has the viscosity of liquid He I and other properties expected of a normal liquid with high zero-point energy at low temperatures. The superfluid component is very different and has virtually no viscosity and no entropy. The two fluids do not interact with each other and each in fact behaves as though the other were not present.

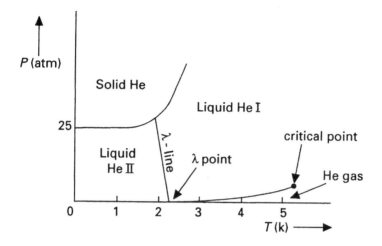

Fig. 8.3. The phase diagram of ⁴He.

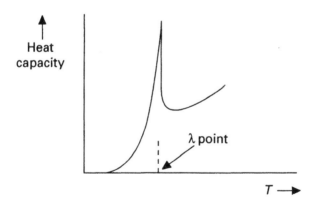

Fig. 8.4. The heat capacity anomaly of liquid ⁴He.

We now consider the relation between the above model and that of a BE gas. To do this we seek to identify the experimental λ-point with the Bose temperature T_B. If we do this we then proceed to suggest that the superfluid component of liquid He II consists of atoms in the energy state $\varepsilon = 0$ into which more and more bosons condense as the temperature falls below T_B; these are the N_0 bosons of equation (8.11). Similarly the normal component of liquid He II is identified as consisting of bosons which are in the excited states above $\varepsilon = 0$; these are the N_{ex} bosons of equation (8.8).

Let us return to equation (8.10). Following Rosser (1982) we treat liquid ^{4}He as an ideal boson gas. The data are:

$$\text{Volume of 1 mole of liquid } ^4\text{He} = 27 \times 10^{-6} \text{ m}^3$$

Therefore

$$\frac{N}{V} = \frac{6 \times 10^{23}}{27 \times 10^{-6}} = 2.2 \times 10^{28} \text{ m}^{-3}$$

Mass of the ^{4}He atom, $m = 6.65 \times 10^{-27}$ kg.
So, from (8.10)

$$T_\text{B} = 3.1 \text{ K}.$$

This compares to the value of 2.17 K for the experimental λ-point. So the Bose temperature of 3.1 K is of the same order of magnitude as the temperature for the transition from He I to He II. The difference between the two temperatures is due in part to the fact that the ideal boson gas model is only an approximation to the actual liquid helium in which atomic interactions play a significant part. So the superfluid properties of liquid He II could be due to a BE type condensation at the λ-point but this is not a complete explanation.

8.5 THE PHOTON GAS: BLACK-BODY RADIATION

The photon is a boson of spin 1.

We start by considering an enclosure of volume V at temperature T. The electromagnetic radiation inside this enclosure may be considered as a gas of photons of different energies. Since photons can be absorbed and re-emitted by the walls of their enclosure the number of photons N is not constant, so that the restriction $\sum_k N_k = N$ (or $\sum_k dN_k = 0$) must not now be applied. So the multiplier α in equation (5.10) is identically zero, that is $e^{-\alpha}$ in equation (8.1a) is now equal to 1. So equations (8.1a,c) become modified to

$$f_k = \frac{N_k}{g_k} = \frac{1}{e^{-\beta\varepsilon_k} - 1} \tag{8.12a}$$

and

$$f(\varepsilon) = \frac{1}{e^{-\beta\varepsilon} - 1} \tag{8.12b}$$

for the number of photons per energy state.

The energy of a photon of frequency ν is $h\nu$ and so the energy distribution of the photons can be expressed in terms of ν. We can rewrite (8.12b) as

$$f(\nu) = \frac{1}{e^{-\beta h\nu} - 1}. \tag{8.12c}$$

We consider the vibrations in the small frequency range ν to $\nu + \mathrm{d}\nu$, each having virtually the same frequency ν and same energy $h\nu$. We now relate $g(\nu)\mathrm{d}\nu$, the number of energy states with frequencies between ν and $\nu + \mathrm{d}\nu$, to $g(k)\mathrm{d}k$ as given by equation (4.7), namely

$$g(k)\mathrm{d}k = \frac{V}{(2\pi)^3} 4\pi k^2 \mathrm{d}k. \tag{8.13}$$

In our photon gas there are two independent directions of polarization of the electromagnetic wave, both being perpendicular to the direction of propagation of the wave. The number of possible modes is therefore doubled since each photon may have either of these directions of polarization. So we multiply the r.h.s. of (8.13) by 2 and write

$$g(k)\mathrm{d}k = \frac{V}{(2\pi)^3} 8\pi k^2 \mathrm{d}k$$

for our number of allowed states in the range k to $k + \mathrm{d}k$.

Now $k = 2\pi\nu/c$ and $\mathrm{d}k = 2\pi\mathrm{d}\nu/c$ where c is the velocity of light. So, rewriting $g(k)\mathrm{d}k$ as $g(\nu)\mathrm{d}\nu$, we readily obtain

$$g(\nu)\mathrm{d}\nu = V 8\pi\nu^2 \mathrm{d}\nu/c^3. \tag{8.14}$$

The energy $u(\nu)\mathrm{d}\nu$ in the range ν to $\nu + \mathrm{d}\nu$ is the number of photons with these ν values times the energy of each, i.e.

$$u(\nu)\mathrm{d}\nu = N(\nu)\mathrm{d}\nu \times h\nu.$$

Also

$$N(\nu)\mathrm{d}\nu = f(\nu)g(\nu)\mathrm{d}\nu$$

(cf. equation (8.4)).

Therefore

$$u(\nu)\mathrm{d}\nu = f(\nu)g(\nu)\mathrm{d}\nu \times h\nu$$

$$= V\frac{8\pi h\nu^3 \mathrm{d}\nu}{c^3} \frac{1}{[\exp(h\nu/k_{\mathrm{B}}T) - 1]} \tag{8.15}$$

which is the Planck radiation formula, giving the spectral distribution of the energy of the radiation inside our constant-temperature enclosure.

If we wish to express Planck's formula in terms of wavelength λ then

$$u(\nu)\mathrm{d}\nu = u(\lambda)\mathrm{d}\lambda$$

and

$$|\mathrm{d}\nu| = \frac{c}{\lambda^2}|\mathrm{d}\lambda|,$$

since $c = \nu\lambda$. Then (8.15) becomes

$$u(\lambda)\mathrm{d}\lambda = V\frac{8\pi hc\,\mathrm{d}\lambda}{\lambda^5[\exp(hc/\lambda k_{\mathrm B}T) - 1]}. \qquad (8.16)$$

The curves of $u(\lambda)$ against λ, for various temperatures, are shown in Fig. 8.5.

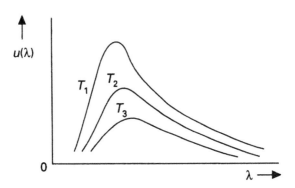

Fig. 8.5. The spectral distribution of the energy of the radiation for temperatures $T_1 > T_2 > T_3$. The area under the curves is proportional to T^4. The values of the wavelength λ_{max} at which $u(\lambda)$ is a maximum lie on the rectangular hyperbola $\lambda_{\mathrm{max}}T = \mathrm{constant}$.

The Planck radiation formula contains all the other laws of radiation. One such law is that the total energy in the radiation is proportional to T^4. This total energy is found by integrating (8.15) over all frequencies or (8.16) over all wavelengths. Thus the total energy per unit volume (or energy density) is

$$\frac{U}{V} = \frac{8\pi h}{c^3}\int_0^\infty \frac{\nu^3\,\mathrm{d}\nu}{[\exp(h\nu/k_{\mathrm B}T) - 1]}$$

$$= \frac{8\pi h}{c^3}\left(\frac{k_{\mathrm B}T}{h}\right)^4\int_0^\infty \frac{x^3\,\mathrm{d}x}{\mathrm{e}^x - 1}$$

on putting $x = h\nu/k_{\mathrm B}T$.

The definite integral on the r.h.s. has the value $\pi^4/15$. So we can write

$$\frac{U}{V} = \sigma T^4$$

where $\sigma = 8\pi^5 k_{\mathrm B}^4/15c^3h^3$.

Now consider equation (8.16). At long wavelengths, $\exp(hc/\lambda k_{\mathrm B}T) \simeq 1 + hc/\lambda k_{\mathrm B}T$ and (8.16) becomes

$$u(\lambda)\mathrm{d}\lambda \simeq V\frac{8\pi k_{\mathrm B}T}{\lambda^4}\mathrm{d}\lambda$$

which is the classical Rayleigh–Jeans formula.

At short wavelengths where $\exp(hc/\lambda k_B T) \gg 1$ (8.16) becomes

$$u(\lambda)\mathrm{d}\lambda = V\left(\frac{8\pi hc}{\lambda^5}\right)e^{-hc/\lambda k_B T}\mathrm{d}\lambda$$

which is Wien's law.

It can also be shown that the wavelength λ_{max} at which $u(\lambda)$ is a maximum satisfies the result $\lambda_{max}T = \text{constant}$, known as Wien's displacement law.

8.6 SUMMARY

In this chapter we have considered the behaviour of an ideal boson gas and the Bose–Einstein condensation. Some of the thermodynamic functions of this gas have been compared with those of an ideal MB gas. We then discussed two applications of the theory, namely, the behaviour of ^{4}He and the properties of a photon gas.

9

Solids

9.1 CLASSICAL THEORY OF THE HEAT CAPACITY OF LATTICES

Consider an assembly of N particles moving in one dimension under harmonic forces. There are $2N$ degrees of freedom; of these N arise from the kinetic energy and N from the potential energy. From the equipartition principle each of these $2N$ degrees of freedom contributes $\frac{1}{2}k_{\mathrm{B}}T$ to the internal energy (see section 6.2). So the total internal energy is

$$U = Nk_{\mathrm{B}}T$$

and the heat capacity is

$$C_V = \mathrm{d}U/\mathrm{d}T = Nk_{\mathrm{B}}$$

which is independent of temperature.

It is easy to extend the above reasoning to a three-dimensional lattice. The molecules of a solid are constrained to oscillate about the lattice sites, each molecule executing harmonic motion. Each has three degrees of freedom and we must assign an energy $k_{\mathrm{B}}T$ to each of these three ($\frac{1}{2}k_{\mathrm{B}}T$ for each of the kinetic and potential energies). Hence the total internal energy is

$$U = 3Nk_{\mathrm{B}}T$$

and the heat capacity is

$$C_V = 3Nk_{\mathrm{B}}$$

(or, per mole, $C_V = 3R$).

This classical result is in good agreement with experiment at sufficiently high temperatures. It is enshrined in the empirical law of Dulong and Petit which tells us that, at temperatures that are high enough, the molar heat capacities at constant volume of all pure substances in the solid state are very nearly equal to $3R$. At low temperatures the observed heat capacities fall to very low values and approach zero as the temperature approaches absolute zero. This is another example where classical theory does not give us the right answer and we must use quantum mechanics to deal with the situation.

9.2 THE QUANTIZED LINEAR OSCILLATOR

Consider an assembly of N identical linear oscillators which we can regard as constituting a one-dimensional crystal lattice. Each linear oscillator is a particle of mass m constrained to move along a straight line (the x-axis) and acted on by a restoring force $-Kx$ proportional to its displacement x from a fixed point. Its equation of motion is

$$m\ddot{x} = -Kx.$$

We treat our assembly as one of N identical quasi-independent localized oscillators (or particles). Its properties form the basis of the theory of the heat capacity of solids. In regarding our oscillators as quasi-independent what we are really saying is this: the interactions between the particles are small enough to ensure that each particle can oscillate virtually independently of the others but are large enough so that there can be energy exchanges between the particles consistent with a given total energy; in other words the oscillators are weakly coupled.

According to quantum mechanics the energy levels of such an oscillator are

$$\varepsilon_j = (j + \frac{1}{2})h\nu$$

where $j = 0, 1, 2, 3, \ldots$ etc. and $\nu = \frac{1}{2\pi}\sqrt{K/m}$ is the classical frequency of the oscillator. The levels are non-degenerate.

The partition function for the oscillator is

$$Z = \sum_j e^{-\varepsilon_j/k_B T}$$

$$= \sum_j \exp\left[-(j + \frac{1}{2})h\nu/k_B T\right]$$

$$= e^{-h\nu/2k_B T} \sum_j \exp(-jh\nu/k_B T).$$

The summation on the r.h.s. of the last equation is equal to

$$1 + e^{-h\nu/k_B T} + e^{-2h\nu/k_B T} + \ldots = \frac{1}{1 - e^{-h\nu/k_B T}}$$

since the series is a simple geometric progression.

Writing $\theta = h\nu/k_B$ we then have

$$Z = \frac{\exp(-\theta/2T)}{1 - \exp(-\theta/T)}, \tag{9.1}$$

where θ is known as the characteristic temperature of the assembly.

From Z we can obtain U, C_V and other quantities. Thus, from (3.17),

$$U = N k_B T^2 \frac{d\ln Z}{dT}$$

$$= \frac{N k_B \theta}{2} + \frac{N k_B \theta}{\exp(\theta/T) - 1} \tag{9.2}$$

and the (U, T) graph is plotted in Fig. 9.1.

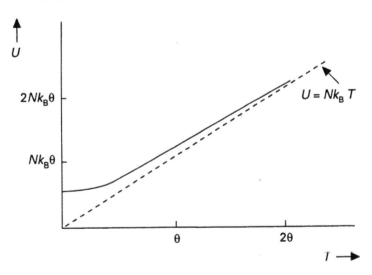

Fig. 9.1. The (U, T) variation for an assembly of harmonic oscillators.

The first term on the r.h.s. of (9.2) represents the zero-point energy; the second is the thermal energy. As $T \to 0$ most of the oscillators are in their lowest energy state of $h\nu/2$ and $U \to Nh\nu/2$, that is, $Nk_B\theta/2$. When $T \gg \theta$, $\theta/T \ll 1$ and $\exp(\theta/T) - 1 \simeq \theta/T$; at this stage

$$U = \frac{Nk_\mathrm{B}\theta}{2} + Nk_\mathrm{B}T$$

$$= Nk_\mathrm{B}T\left[\frac{\theta}{2T} + 1\right]$$

$$\simeq Nk_\mathrm{B}T.$$

So at high temperatures the (U, T) graph approaches the straight dotted line $U = Nk_\mathrm{B}T$. This is in agreement with our discussion in section 9.1.

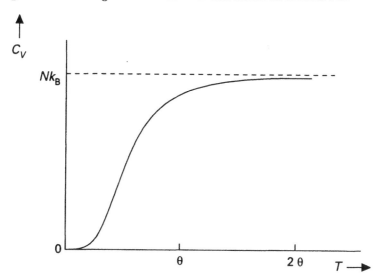

Fig. 9.2. The (C_V, T) variation for an assembly of harmonic oscillators.

The heat capacity of the assembly is

$$C_V = \frac{\mathrm{d}U}{\mathrm{d}T} = Nk_\mathrm{B}\left(\frac{\theta}{T}\right)^2 \frac{\exp(\theta/T)}{[\exp(\theta/T) - 1]^2}$$

and the variation of C_V with T is shown in the graph of Fig. 9.2.

As $T \to 0$, $C_V \to 0$ and when $T \gg \theta$, $C_V \to Nk_\mathrm{B}$ the classical value.

9.3 EINSTEIN'S THEORY OF THE HEAT CAPACITY OF A SOLID

We have seen in section 9.1 that C_V for a solid tends to $3Nk_\mathrm{B}$ at high temperatures and decreases to zero at very low temperatures. The first explanation for this was given by Einstein who considered the atoms of a solid as an assembly of quantized oscillators each vibrating with the same frequency ν. An assembly of N such atoms is equivalent to $3N$ linear oscillators since each atom is free to move in three dimensions. So instead of the result (9.2) for N linear oscillators we get the modified result for $3N$ such oscillators, namely,

$$U = \frac{3Nk_B\theta_E}{2} + \frac{3Nk_B\theta_E}{\exp(\theta_E/T) - 1}$$

where $\theta_E = h\nu/k_B$ is here the Einstein temperature. Hence

$$C_V = \frac{dU}{dT} = 3Nk_B \left(\frac{\theta_E}{T}\right)^2 \frac{\exp(\theta_E/T)}{[\exp(\theta_E/T) - 1]^2}.$$

If C_V is plotted against T/θ_E the general shape of the graph agrees with experiment. When $T \gg \theta_E$, $e^{\theta_E/T} \simeq 1 + \theta_E/T \simeq 1$ and $C_V = 3Nk_B$, the Dulong–Petit value.

As T approaches very low values C_V decreases much more rapidly than do the experimental values. So the Einstein theory is not entirely satisfactory although it represents one of the early successes of quantum mechanics.

9.4 DEBYE'S THEORY

In section 9.2 we discussed an assembly of N linear oscillators and in section 9.3 we extended these ideas to three dimensions when we briefly considered the Einstein theory. There we regarded a crystal of N atoms as an assembly of $3N$ linear oscillators. Debye considered the atoms as a system of coupled oscillators having a continuous spectrum of natural frequencies up to a certain value ν_m; this differs from the simple Einstein theory which assumes that all the atoms oscillate with the same frequency ν. Debye further assumed that the natural frequencies of these atoms would be the same as the frequencies of the possible stationary waves in a crystal if the crystal were a continuous elastic solid.

We have already seen that electromagnetic radiation inside a constant temperature enclosure may be regarded as an assembly of photons of various energies (see section 8.5). Similarly we can regard the energy of the elastic (sound) waves inside a solid medium as being quantized in the form of *phonons*. The energy of a phonon of frequency ν is $h\nu$. For these phonons there is no restriction on the number allowed per energy state, so the assembly of phonons may be considered as a boson gas. If $N(\nu)d\nu$ is the number of phonons with frequencies in the range ν to $\nu + d\nu$, then, from equation (8.5a)

$$N(\nu)d\nu = \frac{g(\nu)d\nu}{e^{h\nu/k_BT} - 1} \tag{9.3}$$

putting $\alpha = 0$ (that is, $A = 1$) as for photons (see section 8.5) and where $g(\nu)d\nu$ is the number of normal modes of vibration (states) for the phonons in the frequency range ν to $\nu + d\nu$.

We now consider how to evaluate $g(\nu)d\nu$. Following Debye, the solid is pictured as a homogeneous, isotropic elastic medium whose atomic structure is disregarded.

The vibrations are regarded as elastic waves propagated in this medium. There are three types of wave involved: one longitudinal and two transverse, all three being propagated in the same direction. The number of allowed modes of vibration in the range ν to $\nu + d\nu$ for all elastic waves with frequencies in *one* direction moving in a solid of volume V is given by

$$g(\nu)d\nu = V\frac{4\pi\nu^2 d\nu}{c^3}$$

where c is the velocity of the waves. This is similar to the result (8.14) except that we divide by 2 for the *one* direction now involved. So when we consider our *three* types of wave we have

$$g(\nu)d\nu = 4\pi V\left(\frac{1}{c_l^3} + \frac{2}{c_t^3}\right)\nu^2 d\nu \tag{9.4}$$

where c_l is the velocity of the longitudinal wave and c_t that of each of the two transverse waves.

The maximum frequency ν_m is determined from the fact that for the total assembly of N atoms comprising the solid there can be only $3N$ independent modes of vibration, that is, $3N$ phonons.

So

$$3N = \int_0^{\nu_m} g(\nu)d\nu = \frac{4\pi V}{3}\left(\frac{1}{c_l^3} + \frac{2}{c_t^3}\right)\nu_m^3.$$

Combining this result with (9.4) we readily obtain

$$g(\nu)d\nu = \frac{9N\nu^2 d\nu}{\nu_m^3}. \tag{9.5}$$

If we now substitute from (9.5) into (9.3) we have

$$N(\nu)d\nu = \frac{9N}{\nu_m^3}\frac{\nu^2 d\nu}{e^{h\nu/k_B T} - 1}$$

for all $\nu \leqslant \nu_m$ and $N(\nu)d\nu = 0$ for $\nu > \nu_m$.

The total energy of the phonons in the frequency range ν to $\nu + d\nu$ is $h\nu N(\nu)d\nu$ and the total energy of the phonon assembly is thus

$$U = \int_0^{\nu_m} h\nu N(\nu)d\nu$$

$$= \frac{9Nh}{\nu_m^3}\int_0^{\nu_m}\frac{\nu^3 d\nu}{e^{h\nu/k_B T} - 1} \tag{9.6}$$

Note that this value of U does not contain the zero-point energy of the assembly; this is of no consequence since the zero-point energy has no effect on the heat capacity.

Differentiating U w.r.t. T we get

$$C_V = \frac{dU}{dT} = \frac{9Nh^2}{\nu_m^3} \frac{1}{k_B T^2} \int_0^{\nu_m} \frac{\nu^4 e^{h\nu/k_B T} d\nu}{(e^{h\nu/k_B T} - 1)^2}.$$

We put $x = h\nu/k_B T$, $x_m = h\nu_m/k_B T = \theta_D/T$, where θ_D is the Debye characteristic temperature. With this change of variable the upper limit of the integral changes from ν_m to $x_m = \theta_D/T$. So we get

$$C_V = 9Nk_B \left(\frac{T}{\theta_D}\right)^3 \int_0^{\theta_D/T} \frac{x^4 e^x}{(e^x - 1)^2} dx. \tag{9.7}$$

For high temperatures when $\theta_D/T \ll 1$ then x is also $\ll 1$ and $e^x \simeq 1 + x \simeq 1$. So from (9.7)

$$C_V \simeq 9Nk_B \left(\frac{T}{\theta_D}\right)^3 \int_0^{\theta_D/T} x^2 dx$$

$$\tag{9.8}$$

$$= 3Nk_B,$$

in agreement with the Dulong and Petit law.

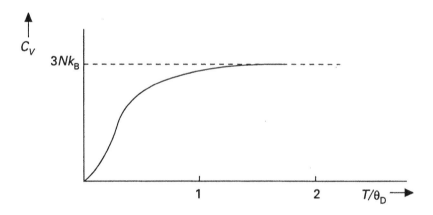

Fig. 9.3. Heat capacity of a solid, according to the Debye theory.

For low temperatures when θ_D/T is very large we let the upper limit in the integral in (9.7) go to infinity.

Now

$$\int_0^\infty \frac{x^4 e^x}{(e^x - 1)^2} dx = \frac{4\pi^4}{15}.$$

So from (9.7)

$$C_V = \frac{12\pi^4}{5} N k_{\mathrm{B}} \left(\frac{T}{\theta_{\mathrm{D}}}\right)^3. \tag{9.9}$$

This equation is valid only when the temperature is lower than about $\theta_{\mathrm{D}}/10$, which for most substances means a temperature below 20 K. It tells us that at low temperatures the heat capacity of a solid varies as T^3 (the Debye T^3 law).

Both results (9.8) and (9.9) are in good agreement with experiment. We see that at high temperatures the Debye theory predicts the value $C_V = 3N k_{\mathrm{B}}$, the classical value. Fig. 9.3 shows how C_V varies with T/θ_{D}. This curve is the same for all monatomic solids, although θ_{D} is not. In fact the values of θ_{D} for various solids are shown in Table 9.1.

Table 9.1. Debye temperatures of some substances

Substance	$\theta_{\mathrm{D}}(\mathrm{K})$
Lead	88
Cadmium	168
Silver	215
Copper	315
Aluminium	398
Iron	453

From the graph of Fig. 9.3 we see that when T/θ_{D} is greater than unity, C_V is nearly equal to the classical value $3N k_{\mathrm{B}}$. When T/θ_{D} is less than unity quantum effects come into play as C_V decreases, eventually reaching zero.

10

Liquids

10.1 INTRODUCTION

The statistical mechanics of gases and solids have been long established but it is only comparatively recently that there has been similar progress in the case of liquids.

During the seventeenth and eighteenth centuries a liquid was regarded as a *continuous medium*. It was at this time that the foundations of hydrodynamics were laid down by Newton, Bernoulli and others, and much of this work remains perfectly valid today. In the early nineteenth century the molecular theory of matter was developed and this led on to the question of the relationships between the properties of individual molecules and the 'bulk' properties of matter. The finite density of matter implies that molecules must repel one another at very short distances while the cohesion of solids implies that there is an attraction between molecules at larger distances apart. Again, when Berthelot in 1850 produced definite experimental evidence that liquids can withstand tensions of at least tens of atmospheres, this showed that the forces between molecules of a liquid also are attractive at larger distances. In 1873 van der Waals showed via his famous equation that intermolecular interactions can, in principle, explain the liquid phase and critical region and these ideas are still valid. The van der Waals isotherm, below a certain temperature, extends into the region of negative pressure thus predicting theoretically that a liquid should sustain a tension. For a further account of this work the reader is referred to the author's (1987) book *Cavitation and tension in liquids.*

The kinetic theory of gases was introduced in the nineteenth century and seems capable of accounting quantitatively for all the properties of gases. The general theory of statistical mechanics was developed by Boltzmann and Gibbs in the later nineteenth century. The advent of quantum theory in the twentieth century had little direct influence on the theory of liquids but it did produce a satisfactory theory of solids, begun by Debye around 1920 (see section 9.4).

In the 1920s the work of the Braggs showed that X-ray scattering could determine the structures of crystalline solids. It was natural to turn to amorphous solids and then to liquids and the appropriate quantity (replacing crystalline structure) was found by Kirkwood, in his pioneering work from 1935 onwards, to be the liquid distribution function $g(r)$ defined in section 10.2(b). Kirkwood's work showed that $g(r)$ and the molecular interaction function $\phi(r)$ can be related by an *integral equation*. The next big step forward occurred in 1957 when a definite check on the integral equation approach was made possible by the use of large computers using Monte Carlo and molecular dynamics methods. A further advance was made in 1958 by the introduction of a new integral equation by Percus and Yevick. For further details see Temperley's (1968) account in *Physics of simple liquids*, Chapter 1 and *Liquids and their properties* by Temperley and Trevena, Chapter 1 (1978).

In this chapter we shall summarize the main attempts which have been made to study liquids. We begin with a description of the gas-like and solid-like approaches to liquids and discuss the concept of the radial distribution function. This is followed by a discussion of intermolecular interactions in liquids. These ideas are then developed by considering (a) the partition function for a dense gas or liquid based on the Mayer theory of imperfect gases and (b) the use of the radial distribution function and of integral equations. Finally there is a brief account of Monte Carlo and molecular dynamics methods.

10.2 APPROACHES TO THE LIQUID PHASE

Broadly speaking, there have been two main approaches to a study of liquids. In one approach a liquid is regarded as a very dense or imperfect gas and in the other it is regarded as a disordered solid. We now consider these two ideas in turn.

(a) The gas-like approach

As we would expect, the similarity between a gas and a liquid is most evident at temperatures near the critical point. We have already met the equation of state $PV = Nk_BT$ for a perfect gas of N 'point' molecules; it was based on the assumption that there are no intermolecular forces between any two molecules and this means that there is no mutual potential energy between them either (see section 10.3). In this case the total internal energy U of the gas is simply the sum of the kinetic energies of the molecules and the result turns out to be $U = 3Nk_BT/2$ as we saw in section 6.6.

Suppose now that we have a *real* gas, rather than the perfect gas envisaged above. If the density of this real gas is low, so that its molecules are far apart for most of the time, we can then assume that there are no effective interactions between the molecules; so at low densities we would expect the equation $PV = Nk_BT$ to be obeyed by a real gas and this is found to be so. However, at higher densities, the molecules of our real gas will be crowded close together and we can no longer ignore the interactions between its molecules.

What can we say about the forces between two such real molecules? The fact that there are cohesive forces holding the molecules together in solids or liquids suggests that forces between molecules are attractive. Yet it is extremely difficult to compress a solid or liquid and so we conclude that, below a certain intermolecular separation, the force between two real molecules becomes strongly repulsive. So for two real molecules we must take account of *repulsions* when they are extremely close together and of *attractions* when they are somewhat further apart.

One way of doing this was proposed by van der Waals in 1873. He modified the equation $PV = RT$ for one mole of gas and wrote it in the form

$$\left(P + \frac{a}{V^2}\right)(V - b) = RT. \tag{10.1}$$

In the van der Waals model the repulsion between two molecules is dealt with by considering the molecules to be rigid spheres of diameter σ rather than as infinitely small mass points. The minimum possible distance between the centres of two molecules will then be σ; at this separation a very strong or infinite repulsion between them occurs. Due to the finite size of the molecules the effective volume available for the molecules to roam in is not V, as for point molecules, but something less, say $(V - b)$. In fact b turns out to be four times the total volume of all the N molecules in the gas, that is, $b = 2N\pi\sigma^3/3$.

Now consider the attractions. When the distance r between two molecules exceeds σ the force between them is an attraction; when r reaches about five molecular diameters, or even less, this attraction is virtually zero. Van der Waals argued that all these attractions were equivalent to an extra internal pressure P' causing the molecules to crowd together and he showed that $P' \propto 1/V^2$ for a constant mass of gas. With his modifications the gas equation therefore assumes the form in (10.1).

Another way of describing the departure from 'perfect gas behaviour' is to consider the virial equation of state which can be written as

$$\frac{PV}{RT} = 1 + \frac{B'}{V} + \frac{C'}{V^2} + \frac{D'}{V^3} + \cdots \tag{10.2}$$

where the virial coefficients B', C', etc. are a measure of how much the behaviour of our imperfect gas departs from that of a perfect gas. This virial equation will be discussed further in section 10.6.

(b) The solid-like approach

Just as the similarity between a liquid and a gas is more evident near the critical point so is that between a liquid and a solid more evident near the triple (or melting) point.

In a crystalline solid at absolute zero, the atoms or molecules are virtually at rest at the lattice sites. These sites form a definite regular geometrical pattern and because this regular arrangement, or 'order', exists throughout the whole inside region, the solid is said to possess long-range order. If we choose any internal molecule as a reference point the position of any other molecule, however far removed, will bear some definite geometrical correlation to the position of this reference molecule. Each molecule will also have a certain number, z, of nearest neighbours; z is called the coordination number of the lattice. When we heat up the solid above absolute zero the molecules start to vibrate more and more as the temperature rises but these vibrations still occur about a well-defined pattern of sites so that long-range order is still preserved.

Now consider what happens as the temperature is raised still further. A struggle occurs between the ordering influence of the intermolecular forces and the disordering influence of the increasing thermal motion of the molecules. When the solid eventually melts into a liquid the original long-range order is destroyed but a good deal of local or short-range order remains. By this we mean that the molecular arrangement near a chosen central molecule in the liquid will be a fairly regular one but when we move out to distances of 30 Å = 3 nm or so the positions of the molecules there will bear no real spatial geometrical relations to the position of our reference molecule. In other words, there is short-range order just around any given molecule but no long-range order throughout the liquid.

This local order is described by the *radial distribution function*, $g(r)$. We shall confine our attention to a simple liquid, such as liquid argon, in which the atoms or molecules are spherically symmetrical. Each molecule can then be regarded as a small sphere of diameter σ and the force between two molecules will depend only on the separation of their centres.

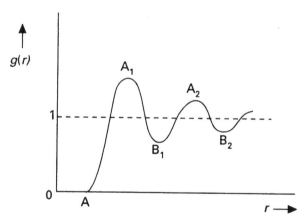

Fig. 10.1. The radial distribution function describing the local order
in a simple liquid.

Let us take one molecule of centre O and consider the surrounding environment as viewed by an observer at O. The number of molecules per unit volume at a distance r from O will be denoted by $\rho(r)$, the *number density*. As we move out along any straight line from O, $\rho(r)$ will vary with r; owing to the nature of short-range order $\rho(r)$ will sometimes be more, sometimes less, than the mean number density ρ_0 throughout the whole liquid where

$$\rho_0 = \frac{\text{Total number of molecules in the liquid}}{\text{Total volume of the liquid}}$$

In other words the ratio $g(r) = \rho(r)/\rho_0$ can be greater than, less than or equal to unity. The function $g(r)$ is the radial distribution function and the graph of $g(r)$ against r is of the general shape shown in Fig. 10.1.

The curve starts from zero at the point A where the distance OA is equal to the diameter of a molecule; a little thought will show that this must be so. For large r, the value of $\rho(r)$ will tend to ρ_0, that is, $g(r)$ tends to unity. For distances of a few molecular diameters $g(r)$ departs from unity and this departure is a measure of the short-range order in the liquid around O. The first maximum A_1 corresponds to the inner shell of O's nearest neighbours, A_2 to the second shell of neighbours and so on; the minima B_1, B_2, etc. correspond to the 'emptier' regions between these shells. For example, for liquid argon at 84 K the values of r at the first and second maxima are about 4.0 and 7.0 Å.

It must be emphasized that both $g(r)$ and $\rho(r)$ for any value of r represent the mean values averaged over time at that distance; large statistical fluctuations from these values will occur owing to the thermal motion of the molecules. These functions have been obtained from experiments involving the diffraction of X-rays and neutrons by liquids. Furthermore, as we shall see in section 10.7, quantities such as the internal energy and the equation of state for a liquid can be calculated directly from $\rho(r)$.

10.3 THE GENERAL NATURE OF INTERMOLECULAR INTERACTIONS IN LIQUIDS

Consider a simple liquid consisting of N identical spherical molecules. First we consider just two of these molecules, labelled i and j, and ignore the presence of the others. Let r be the distance between their centres. Then the mutual potential energy $\phi(r)$ of the two molecules varies with r as shown in the graph of Fig. 10.2.

The value of ϕ is zero at $r = \sigma$. We define σ as the effective diameter of each molecule. At any stage the force between the molecules is

$$F = -\mathrm{d}\phi/\mathrm{d}r.$$

For $r < \sigma$ the curve is practically vertical which implies a very large or infinite repulsion between the molecules. As r increases beyond σ, ϕ reaches a minimum of $-\varepsilon$ at $r = r_0$ and at this point the repulsion changes to an attraction; for $r > r_0$

the (ϕ, r) curve quickly tends to the r-axis over a distance of only a few molecular diameters.

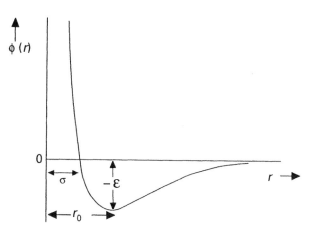

Fig. 10.2. Variation of $\phi(r)$ with r.

Now consider the whole liquid. Let the molecules be labelled 1 to N. The number of distinct pairs of molecules ij is $NC_2 = N(N-1)/2$. If the mutual potential energy of molecules i and j at distance r_{ij} apart is $\phi(r_{ij})$ and if we assume that the molecules interact in pairs only then the total potential energy Φ of the assembly will be

$$\Phi = \sum_{i>j} \phi(r_{ij}) \tag{10.3}$$

where $\displaystyle\sum_{i>j}$ implies that each $\phi(r_{ij})$ is not counted twice and the summation in (10.3) contains a total of $N(N-1)/2$ terms.

For a fuller description see Temperley and Trevena's *Liquids and their properties* (1978), Chapter 3.

10.4 THE 'RIGID-SPHERE' AND OTHER SIMPLE FUNCTIONS

We now mention three simpler forms of $\phi(r)$ which have been used in work on liquids.

In the first of these each molecule is regarded as being a non-attracting rigid sphere of diameter σ. This 'rigid-sphere' potential energy is then defined by $\phi(r) = \infty$ for $r < \sigma$ and $\phi(r) = 0$ for $r > \sigma$. A more realistic approximation is that in which the molecules are rigid spheres of diameter σ acting on each other with weak attractive forces. In this case $\phi(r) = \infty$ for $r < \sigma$ and $\phi(r) = -A/r^6$ for

$r > \sigma$. A third function sometimes used is the 'square-well' potential defined by

$$\phi(r) = \infty, \quad r < \sigma;$$
$$\phi(r) = -\varepsilon, \quad \sigma < r < \sigma_1;$$
$$\phi(r) = 0, \quad r > \sigma_1,$$

where σ_1 is a length greater than σ and ε is an energy. σ_1/σ is usually taken to be about 1.5.

10.5 THE PARTITION FUNCTION FOR AN IMPERFECT GAS OR LIQUID

We have already seen in section 6.6 that the partition function for a perfect gas of N molecules is given by

$$Z_{\mathrm{W}} = Z^N/N! \tag{10.4}$$

where $Z = V(2\pi m k_{\mathrm{B}} T/h^2)^{3/2}$ is the partition function for one molecule. This result takes into account the translational kinetic energies of the molecules only and it is assumed that there are no intermolecular interactions.

Next let our assembly be either an imperfect (dense) gas or a liquid, consisting of N identical molecules. The interactions between the molecules must now be taken into account in addition to their translational energies. If we assume that the molecules interact in pairs only, the total potential energy Φ of the assembly is given by (10.3), that is,

$$\Phi = \sum_{i>j} \phi(r_{ij}).$$

It can then be shown that the partition function for the whole assembly is

$$Z'_{\mathrm{W}} = \frac{1}{N!} \left(\frac{2\pi m k_{\mathrm{B}} T}{h^2} \right)^{3N/2} \int \ldots \int \exp(-\Phi/k_{\mathrm{B}} T) \mathrm{d}\tau_1 \mathrm{d}\tau_2 \ldots \mathrm{d}\tau_N \tag{10.5}$$

where the integral is taken over $\mathrm{d}\tau_i = \mathrm{d}x_i \mathrm{d}y_i \mathrm{d}z_i$ for all values of i from 1 to N, that is, for all positions of all the molecules within our volume V.

We shall write (10.5) as

$$Z'_{\mathrm{W}} = \left(\frac{2\pi m k_{\mathrm{B}} T}{h^2} \right)^{3N/2} \frac{Q_N}{N!}$$

where

$$Q_N = \int \ldots \int \exp(-\Phi/k_\mathrm{B}T)\mathrm{d}\tau_1 \ldots \mathrm{d}\tau_N. \tag{10.6}$$

Q_N is the *configurational integral* and is simply the Boltzmann factor associated with the intermolecular energy, Φ, integrated over all possible positions of the N molecules in an enclosure of volume V. The crux of the problem is the evaluation of Q_N; this will be discussed in the next section. Once Q_N is obtained Z'_W will also be known and the various thermodynamic properties can be derived.

10.6 THE MAYERS' VIRIAL EXPANSION

For a full account of the Mayers' work relating the properties of imperfect gases to the intermolecular function $\phi(r_{ij})$ or ϕ_{ij} the reader must turn to the Mayers' book *Statistical mechanics* (1977) or to similar advanced treatises. Their method can, in principle, be extended to describe the liquid phase itself. Their speculations about the critical region are not now accepted but this does not invalidate the basic soundness of their approach to the imperfect gas and liquid problems.

To evaluate Q_N we first define the Mayer function $f(r_{ij})$ or f_{ij} by the equation

$$\exp(-\phi_{ij}/k_\mathrm{B}T) = 1 + f_{ij}. \tag{10.7}$$

The graph showing the variation of ϕ_{ij} with r_{ij} is that in Fig. 10.2; that showing how f_{ij} varies with r_{ij} is shown in Fig. 10.3.

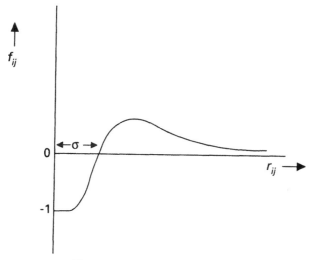

Fig. 10.3. Variation of f_{ij} with r_{ij}.

The important property of f_{ij}, for our present discussion, is that f_{ij} is effectively zero when r_{ij} is large compared with σ; in other words f_{ij} is zero unless our two molecules are very close together.

We now express the integrand in (10.6) in terms of the f-functions. We have

$$\exp(-\Phi/k_\mathrm{B}T) = \exp\left(-\sum_{i>j}\phi_{ij}/k_\mathrm{B}T\right) = \prod_{i>j}(1+f_{ij}) \tag{10.8}$$

so that

$$Q_N = \int\ldots\int\prod_{i>j}(1+f_{ij})\mathrm{d}\tau_1\ldots\mathrm{d}\tau_N. \tag{10.9}$$

The integrand on the r.h.s. of (10.9) can be written as

$$\text{Integrand} = 1 + \sum_{i>j}f_{ij} + f_{12}f_{23} + \cdots \tag{10.10}$$

This has to be integrated over all positions of all the molecules. Integrating the first term of unity gives

$$\int_{(N)}\mathrm{d}\tau_1\mathrm{d}\tau_2\ldots\mathrm{d}\tau_N = V^N$$

which is what we should get for a perfect gas (see equation (10.4)).

Now consider the second term $\sum_{i>j}f_{ij}$. The f-function f_{12} in this summation when integrated gives us

$$\int\int f_{12}\mathrm{d}\tau_1\mathrm{d}\tau_2\int_{(N-2)}\mathrm{d}\tau_3\ldots\mathrm{d}\tau_N = V^{N-2}\int\int f_{12}\mathrm{d}\tau_1\mathrm{d}\tau_2. \tag{10.11}$$

A similar result holds for each of the f_{ij}'s in the summation and because the function f_{ij} is the same for all pairs of molecules we have $N(N-1)/2$ terms of type (10.11), all numerically the same. Each f_{ij} represents an f-factor connecting two molecules i and j, thus

which can be regarded as a group or cluster of two molecules.

Similarly the next terms in (10.10) such as $f_{12}f_{23}$ are involved with clusters of three molecules bounded by two f-factors f_{12} and f_{23}, and so on.

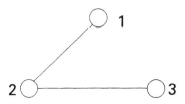

We therefore see that the various terms in (10.10) can be regarded as describing the departure of the assembly from perfect gas behaviour. Since the function f_{ij} is only appreciable when molecules i and j are close together we are concerned with the situation where we have clusters of $2, 3, 4, \ldots, l$, etc. molecules in which all the individual molecules are close together. These various clusters give rise, in Q_N, to integrals of products of f-factors (see equation (10.9)). These higher terms in the expansion of (10.9) occur as cluster integrals of order $2, 3, 4, \ldots, l$, etc. and are written as $b_2, b_3, b_4, \ldots, b_l$ etc. Thus

$$b_2 = \frac{1}{2!V} \int \int f_{12} \mathrm{d}\tau_1 \mathrm{d}\tau_2$$

etc.

As we might expect, the higher-order cluster integrals become relatively more and more important in determining the equation of state as the density is increased. The full argument can be made mathematically rigorous and specifically we find that

$$\frac{P}{k_\mathrm{B} T} = \sum_l b_l \mathbf{Z}^l \tag{10.12}$$

where $\mathbf{Z}$, an auxiliary variable, satisfies the relation

$$N/V = \sum_l l b_l \mathbf{Z}^l. \tag{10.13}$$

The value of b_1 is defined to be unity. For very low densities, both series (10.12) and (10.13) reduce to their first terms and we have the perfect gas law. For somewhat higher densities, we have in turn the corrections to the perfect gas law associated with the interactions of 2, 3, 4 and so on molecules with one another. Mathematically, by eliminating the variable $\mathbf{Z}$ between (10.12) and (10.13) we obtain the virial series

$$\frac{P}{k_\mathrm{B} T} = N/V + B[N/V]^2 + C[N/V]^3 + D[N/V]^4 + \ldots \tag{10.14}$$

which is an expansion for pressure in ascending powers of N/V and this expansion is certainly valid if the series (10.12) and (10.13) are convergent. This is true at low enough densities for a wide range of possible interaction functions. We conclude that, at sufficiently low densities, the departure of a gas from perfection should be described by (10.14). In this equation it can be shown from the Mayers' theory that the second virial coefficient B is related to the Mayer function f by

$$B = -2\pi N \int_0^\infty r^2 f(r) \mathrm{d}r \tag{10.15}$$

and the higher virial coefficients C, D etc. are likewise expressible as multiple integrals of products of f-functions.

The virial series, in which virial coefficients beyond the fifth are neglected, gives reasonable values for the properties of gases below the critical density. The

Mayers' method, to some extent, also explains the phenomenon of condensation. At liquid densities, however, the virial expansion fails to converge and so cannot give meaningful results.

10.7 THE USE OF THE RADIAL DISTRIBUTION FUNCTION

It is possible to derive the total internal energy and the equation of state of our liquid of N molecules in terms of the radial distribution function $g(r)$ or, what is really the same, the number density $\rho(r)$ (see section 10.2(b)).

First we consider the internal energy U. We choose a reference molecule i as our origin and then work out the mutual potential energy between i and all the molecules in the spherical shell between r and $r + dr$ from the origin. In this shell there are $\rho(r)4\pi r^2 dr$ molecules, on average. This gives the total potential energy of i as

$$\int_0^\infty \phi(r)\rho(r)4\pi r^2 dr$$

since we must integrate over all values of r. We may take the upper limit of our integral as infinity because $\phi(r)$ tends quickly to zero with distance.

To find the total potential energy of the liquid we must repeat this procedure for each of the N molecules in our liquid and then divide by two to prevent each pair of molecules from being counted twice. Finally, to obtain the total internal energy U we must also include the kinetic energy $3Nk_BT/2$. Thus we have

$$U = \frac{3Nk_BT}{2} + \frac{N}{2} \int_0^\infty \phi(r)\rho(r)4\pi r^2 dr. \qquad (10.16)$$

We next consider the equation of state using a similar argument. This is given by Clausius' virial theorem for molecules with central forces, namely,

$$PV = Nk_BT + \frac{1}{3}\overline{\sum_{\text{pairs}} F(r_{ij})r_{ij}}. \qquad (10.17)$$

What this tells us is that we need to multiply the central force $F(r_{ij})$ between any pair of molecules i and j by their distance apart r_{ij}; we then sum these products over *all* pairs in the liquid, take the average of this sum, as measured at various times, and finally substitute the result into equation (10.17). For a derivation of equation (10.17) the reader is referred to, for, example, Pryde's *The liquid state* (1966), Chapter 5.

Again choosing i as our central reference molecule and j as a molecule in our spherical shell of radius $r = r_{ij}$, then, from section 10.3,

$$F(r_{ij}) = F(r) = -d\phi(r)/dr.$$

Since there are $\rho(r)4\pi r^2 dr$ molecules in our shell the contribution to the sum on the r.h.s. of (10.17) is

$$\frac{1}{3}r\left(-\frac{d\phi(r)}{dr}\right)\rho(r)4\pi r^2 dr.$$

Using the same argument as before we integrate this expression and then multiply by $N/2$. Thus (10.17) becomes

$$PV = Nk_BT - \frac{N}{6}\int_0^\infty \frac{d\phi(r)}{dr}\rho(r)4\pi r^3 dr. \qquad (10.18)$$

Having derived expressions for the internal energy and equation of state in terms of $\phi(r)$ and $\rho(r)$ it is possible, in principle, to calculate U, P and other equilibrium thermodynamic properties. To do this we assume a given form for $\phi(r)$. However, obtaining a value for $\rho(r)$ presents difficulties; it can be obtained from diffraction experiments but the difficulty is that such values of $\rho(r)$ cannot be measured accurately and the pressure turns out to be very sensitive to the value of $\rho(r)$. So an attempt had to be made to calculate $\rho(r)$ from first principles. Kirkwood saw that it was mathematically possible to do this. He showed in 1935 that $\phi(r)$ and $\rho(r)$ can be related by an *integral equation*. This equation has never been solved exactly but Kirkwood and his students were able to solve various approximate versions of it. These amounted to the first *a priori* calculations of the equation of state of a liquid.

The mathematical details of this work are very complicated and further modifications have been tried by other workers. The form of integral equation which works best is due to Percus and Yevick (1958) and $\rho(r)$ can now be predicted theoretically to an accuracy comparable to that with which it can be measured. Once $\rho(r)$ is known in addition to $\phi(r)$ all the equilibrium properties are known also.

One other point is worth mentioning. The distribution function for any liquid, no matter what the intermolecular interaction is like, is very similar to that one would expect if the molecules were rigid spheres (see section 10.4). In other words, the form of $\rho(r)$ is determined largely by geometrical considerations while the fact that real molecules attract one another and are not absolutely 'rigid' spheres does not affect it very much. In fact, the rigid-sphere liquid is much closer in structure to a real liquid than are either a modified solid or a dense gas which the older models took as starting points.

10.8 MONTE CARLO AND MOLECULAR DYNAMICS METHODS

These methods consist of explicit calculations of various properties of small assemblies of the order of 100 molecules and their execution requires a large and fast computer. The properties of these very small assemblies turn out to be not very different from those of larger ones; thus, by working with assemblies of different sizes we can extrapolate the results to real assemblies containing very large numbers of molecules.

In the Monte Carlo method we start off with the assembly in some standard configuration and then generate a subsequent sequence of configurations by moving the molecules one at a time, thus simulating the effect of thermal agitation. We then average the particular property in which we are interested over all the chosen configurations. Calculations for such small assemblies of rigid spheres and of molecules interacting with other forms of $\phi(r)$ have been made. At different stages in the history of the assembly we can calculate such things as the radial distribution function, the mean kinetic energy per molecule or the pressure. The configurational partition function Q_N can also be estimated.

The molecular dynamics method is more ambitious. We begin, as before, with a number of molecules in some given configuration in a box; the molecules are then started off with equal speeds but with random directions of motion. The resulting changes in the assembly with time are followed by solving the individual equations of motion of all the molecules. 'Periodic' boundary conditions are chosen so that a molecule passing out of one side of the box re-enters with the same velocity through the opposite side, thus ensuring that the number of molecules in the box remains constant. The equilibrium value of any particular property is found by time averaging and quantities such as the pressure and radial distribution can be found. Alder and Wainwright (1957, 1959) used these methods to study an assembly of 32 rigid spheres and one of 108 molecules with square-well potentials. The method also enables one to study the solidification process and other non-equilibrium effects.

One of the most significant discoveries was made independently in 1957 by each of these two methods. This was that, at a density of 2/3 that of close-packed density, an assembly of rigid spheres would make a transition from a disordered liquid-like configuration to an ordered solid-like configuration. (In close-packed density the spheres are as closely packed as is geometrically possible, each with twelve nearest neighbours.) This transition corresponds to the solidification of a real gas or liquid and occurs even though there are no attractive forces between the molecules. The existence of such a transition for a rigid sphere assembly had been predicted by Kirkwood and Monroe but the prediction had remained suspect because of the drastic approximations made in the theory. Furthermore, many people found it difficult to accept the idea that purely repulsive interactions could bring about the appearance of an ordered structure like a lattice.

For further details of the Monte Carlo and molecular dynamics studies the reader is referred to Temperley et al.'s *The physics of simple liquids* (1968), Chapters 4 and 5 and to the author's *The liquid phase* (1975), Chapter 5.

10.9 A FINAL SUMMARY

In this chapter we have seen that the application of statistical mechanics to liquids involves an assumed form for the intermolecular energy function $\phi(r)$. Various forms of $\phi(r)$ have been discussed. With the appropriate form of $\phi(r)$ a liquid can be treated as an imperfect gas, using the elegant Mayer theory, or as a quasi-crystalline solid using the radial distribution function.

The Monte Carlo and molecular dynamics methods also involve some assumption about $\phi(r)$. These methods are able to simulate the molecular situations in a gas, liquid and solid and also the transition from one of the phases to another.

11

Other related topics

In this chapter we shall consider a few topics which perhaps do not fit naturally into any of the preceding chapters.

11.1 THE RESULTS $P = 2U/3V$ FOR AN IDEAL MONATOMIC GAS AND $P = U/3V$ FOR A PHOTON GAS

For a Maxwell–Boltzmann gas we have the results $PV = Nk_\mathrm{B}T$ and $U = \frac{3}{2}Nk_\mathrm{B}T$ (see section 6.6). So the relation $P = 2U/3V$ follows at once.

This relation, however, is valid for any ideal monatomic gas irrespective of the statistics obeyed by the gas. Since there are no interactions between the particles we need only consider the translational motion of a particle which is given by equation (4.3), namely,

$$\varepsilon_j = n_j^2 h^2 / 8mV^{2/3}.$$

Each value of $\varepsilon_j \propto V^{-2/3}$, that is $\varepsilon_j = C_j V^{-2/3}$.

Also from thermodynamics (see Appendix 4),

$$P = -\left(\frac{\partial U}{\partial V}\right)_S$$

and

$$U = \sum_j n_j \varepsilon_j.$$

Now S depends on the n_js only and the ε_js depend on V, and so, for a constant S the set of n_js does not change. Hence

$$P = -\sum_j n_j \frac{\partial \varepsilon_j}{\partial V}.$$

Now

$$\frac{\partial \varepsilon_j}{\partial V} = -\frac{2}{3} C_j V^{-5/3} = -\frac{2\varepsilon_j}{3V}$$

and so

$$P = \frac{2}{3V} \sum_j n_j \varepsilon_j = \frac{2U}{3V}. \tag{11.1}$$

This result (11.1) holds for MB, FD and BE gases of material particles but not for a gas of photons as we shall now show.

For a photon

$$\varepsilon = h\nu = hc/\lambda = hck/2\pi.$$

From section 4.4 we know that k is proportional to $1/a$, where a is the side of our cubical enclosure of volume $V = a^3$. So, since $\varepsilon \propto k$ from the last equation it follows that $\varepsilon \propto V^{-1/3}$ for photons rather than $\propto V^{-2/3}$ as for material particles. Hence, if we go through a procedure such as that above for material particles we see that, for photons,

$$P = \frac{U}{3V}. \tag{11.2}$$

So the pressure of the radiation is one-third of the energy per unit volume (cf. section 8.5).

11.2 AN ASSEMBLY OF PARTICLES, EACH WITH TWO ENERGY LEVELS; THE SCHOTTKY ANOMALY

Consider an assembly of N localized quasi-independent particles, each of which can be in either of two non-degenerate energy levels (states) ε_0 and ε_1. From (3.4) the partition function of a particle is

$$Z = \sum_j e^{-\varepsilon_j/k_B T}$$

$$= e^{-\varepsilon_0/k_B T} + e^{-\varepsilon_1/k_B T}$$

$$= e^{-\varepsilon_0/k_B T} \left[1 + e^{-\varepsilon/k_B T} \right]$$

where $\varepsilon = \varepsilon_1 - \varepsilon_0$ is the energy difference between the levels. If n_0, n_1 are the numbers of particles in the levels, then $n_0 + n_1 = N$.

From (3.13a) the Helmholtz free energy of the assembly is

$$
\begin{aligned}
F &= -Nk_{\mathrm{B}}T \ln Z \\
&= -Nk_{\mathrm{B}}T \ln e^{-\varepsilon_0/k_{\mathrm{B}}T} \left[1 + e^{-\varepsilon/k_{\mathrm{B}}T} \right]
\end{aligned}
\tag{11.3}
$$

and since, in general $n_j = (Ne^{-\varepsilon_j/k_{\mathrm{B}}T})/Z$ (equation (3.10b)) we have

$$
\begin{aligned}
n_0 &= \frac{N}{Z} e^{-\varepsilon_0/k_{\mathrm{B}}T} = \frac{N}{1 + e^{-\varepsilon/k_{\mathrm{B}}T}} \qquad &\text{(a)} \\
n_1 &= \frac{N}{Z} e^{-\varepsilon_1/k_{\mathrm{B}}T} = \frac{N}{e^{\varepsilon/k_{\mathrm{B}}T} + 1} \qquad &\text{(b)}
\end{aligned}
\tag{11.4}
$$

At high temperatures $(T \to \infty)$, $k_{\mathrm{B}}T \gg \varepsilon$ and the quantities $e^{-\varepsilon/k_{\mathrm{B}}T}$ and $e^{\varepsilon/k_{\mathrm{B}}T}$ both tend to 1 and so n_0 and n_1 both tend to $N/2$. This is not surprising; since $k_{\mathrm{B}}T$, the thermal energy, greatly exceeds the energy gap ε between the two levels the probability of occupation of each level is the same. Conversely, at low temperatures $(T \to 0)$, $e^{-\varepsilon/k_{\mathrm{B}}T} \to 0$ and $e^{\varepsilon/k_{\mathrm{B}}T} \to \infty$. So $n_0 \to N$ and $n_1 \to 0$. Thus at $T = 0$ all the particles are in their lowest energy states and the assembly is completely ordered $(\Omega = 1)$.

Next consider the internal energy U. We have

$$
U = n_0\varepsilon_0 + n_1\varepsilon_1
$$

$$
= N\varepsilon_0 + n_1\varepsilon
\tag{11.5}
$$

$$
= U_0 + n_1\varepsilon
$$

where U_0 is the zero-point energy. As $T \to 0$, $U \to U_0$ with all the particles in their lowest energy states. At high temperatures, when $k_{\mathrm{B}}T \gg \varepsilon$, $n_1\varepsilon \to N\varepsilon/2$ and $U \to U_0 + N\varepsilon/2$. From (11.4b) and (11.5)

$$
U = U_0 + \frac{N\varepsilon}{e^{\varepsilon/k_{\mathrm{B}}T} + 1}
$$

$$
= U_0 + \frac{N\varepsilon}{e^{\theta/T} + 1} .
$$

where $\theta = \varepsilon/k_{\mathrm{B}}$ is the characteristic temperature of the assembly.

With the volume (and, therefore, the energy levels) kept constant the heat capacity is

$$
C_V = \left(\frac{\partial U}{\partial T} \right)_V = Nk_{\mathrm{B}} \left(\frac{\theta}{T} \right)^2 \frac{e^{-\theta/T}}{(1 + e^{-\theta/T})^2} .
$$

We note that throughout the above discussion the important ratio has been that of the two energies ε and $k_B T$; we have also written this as θ/T, the ratio of two temperatures.

The graph of C_V against T is shown in Fig. 11.1. From it we see that C_V

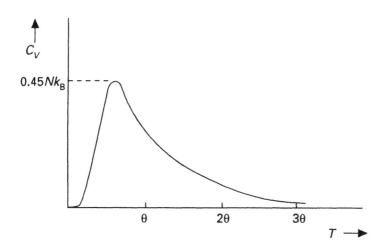

Fig. 11.1. Heat capacity as a function of temperature, showing the anomalous Schottky peak.

has a maximum of about $0.45Nk_B$ in the region where $T \sim \theta$ and is zero for $T = 0$ and $T = \infty$. Heat capacities of this type are associated with a two-level assembly and are known as Schottky anomalies. This sort of heat capacity is found in some paramagnetic salts at low temperatures. A Schottky peak in the heat capacity occurs also in cases like O_2 and NO when there are electronic energy levels fairly near the ground level; we met this in section 6.10(d) when we discussed the electronic heat capacity of a diatomic gas.

Another well-known example of a two-level case is the Zeeman effect in spectroscopy in which a single energy level in an atom splits into two levels when the atom is subjected to a magnetic field.

Next consider the entropy. Using the ideas in section 2.4 we have for this two-level assembly

$$\Omega = \frac{N!}{n_0! n_1!}$$

and

$$S = k_B \ln \Omega$$

$$= k_B[N \ln N - n_0 \ln n_0 - n_1 \ln n_1]. \tag{11.6}$$

Consider first the two extreme cases $T = 0$ and $T \to \infty$. At $T = 0$, $n_0 = N$, $n_1 = 0$ and so $S = 0$ and $\Omega = 1$; that is, there is only one way of arranging the completely ordered configuration with all the particles in the ground state. As $T \to \infty$, $n_0 = n_1 = N/2$ and so, from (11.6),

$$S = N k_B \ln 2.$$

At these high temperatures we can rewrite this as

$$S = k_B \ln 2^N$$

so that

$$\Omega = 2^N.$$

This follows because, as we saw above, the probability of occupation of each of the two levels ε_0 and ε_1 is the same at high temperatures. As $T \to \infty$, S reaches its maximum value and we have a state of complete disorder.

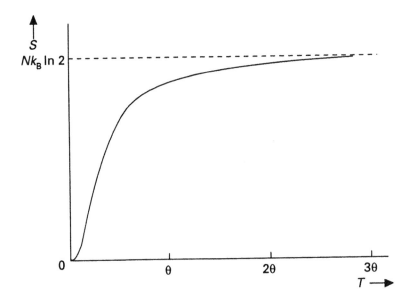

Fig. 11.2. Variation of entropy S with temperature T occurring when there is a Schottky anomaly.

At any intermediate temperature we can calculate S as follows. Equation (11.3) may be written as

$$F = N\varepsilon_0 - Nk_BT\ln\left[1 + e^{-\theta/T}\right]$$

and then

$$S = -\left(\frac{\partial F}{\partial T}\right)_{V,N}$$

$$= Nk_B\left[\ln\left(1 + e^{-\theta/T}\right) + \frac{\theta}{T}\frac{e^{-\theta/T}}{(1 + e^{-\theta/T})}\right].$$

This variation of S with T is sketched in Fig. 11.2.

11.3 THE ENTROPY OF MIXING OF TWO PERFECT GASES

We start by reminding ourselves of the Sackur–Tetrode result (equation (6.20)) for the entropy of a perfect gas, namely,

$$S = Nk_B\left[\ln(V/N) + \frac{3}{2}\ln T + \frac{3}{2}\ln\left(\frac{2\pi mk_B}{h^2}\right) + \frac{5}{2}\right]$$

which we can write as

$$S = Nk_B[\ln(V/N) + C]. \tag{11.7}$$

Referring to Fig. 11.3 we consider two equal volumes V each containing N atoms of the same perfect monatomic gas. The gas in the two volumes (1) and (2) are in thermal equilibrium at a temperature T and are separated by a wall AB which is a good conductor of heat.

From (11.7) the entropy of the gas in each of (1) and (2) is

$$S_1 = S_2 = Nk_B[\ln(V/N) + C].$$

So the total initial entropy of the two separate volumes of gas is

$$S_i = S_1 + S_2 = 2Nk_B[\ln(V/N) + C].$$

If we remove the wall AB, T being kept constant, we shall have $2N$ atoms of the same gas in a volume $2V$ and from (11.7) the total entropy is

$$S_f = 2Nk_B[\ln(2V/2N) + C]$$

which is the same as S_i. So there is no change of entropy on mixing.

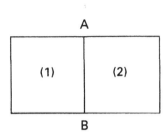

Fig. 11.3. Diagram illustrating the entropy of mixing of two perfect gases.

Next, however, consider the case when volumes (1) and (2) contain *different* inert monatomic gases 1 and 2 (for example, argon and neon) at the same temperature T. Let there be N atoms of type 1 and N of type 2 in the compartments (1) and (2). This means that in equation (11.7) the values of C for the two gases are different since the values of m (m_1 and m_2) are different. Hence, from (11.7), the initial entropies of the two gases are

$$S_1 = Nk_{\mathrm{B}}[\ln(V/N) + C_1]$$

and

$$S_2 = Nk_{\mathrm{B}}[\ln(V/N) + C_2].$$

Thus the total initial entropy is

$$S_{\mathrm{i}} = S_1 + S_2 = 2Nk_{\mathrm{B}}\ln(V/N) + Nk_{\mathrm{B}}[C_1 + C_2].$$

On removing the wall AB the 'playground' for each atom of both gases increases from V to $2V$; hence the final entropy is

$$\begin{aligned} S_{\mathrm{f}} &= Nk_{\mathrm{B}}[\ln(2V/N) + C_1] + Nk_{\mathrm{B}}[\ln(2V/N) + C_2] \\ &= 2Nk_{\mathrm{B}}\ln(2V/N) + Nk_{\mathrm{B}}[C_1 + C_2]. \end{aligned}$$

So there is an entropy of mixing of

$$\Delta S = S_{\mathrm{f}} - S_{\mathrm{i}} = 2Nk_{\mathrm{B}}\ln 2.$$

This expression for the entropy of mixing of two different inert perfect gases is the same, whichever two gases are involved. When the wall AB is removed, diffusion of the gases occurs; there is an increase in disorder and therefore in entropy. However, when the two gases are the same, removal of the wall causes no change in disorder; the idea of diffusion is now meaningless and there is no change in entropy.

12

Worked examples

This chapter will be devoted to a selection of worked examples relating to the various previous topics covered in this book.

EXAMPLE 12.1

An assembly consists of five distinguishable (localized) particles each of which may have an energy $0, \varepsilon, 2\varepsilon$, etc. The internal energy U is 5ε. Show that there are 7 distributions and a total of 126 microstates.

We use the theory given in section 2.3 but here for $N = 5$ and write down the distributions as in Table 2.3. They are

$$
\begin{array}{cccccc}
\text{I} & 5\varepsilon, & 0, & 0, & 0, & 0 \\
\text{II} & 4\varepsilon, & \varepsilon, & 0, & 0, & 0 \\
\text{III} & 3\varepsilon, & 2\varepsilon, & 0, & 0, & 0 \\
\text{IV} & 3\varepsilon, & \varepsilon, & \varepsilon, & 0, & 0 \\
\text{V} & 2\varepsilon, & 2\varepsilon, & \varepsilon, & 0, & 0 \\
\text{VI} & 2\varepsilon, & \varepsilon, & \varepsilon, & \varepsilon, & 0 \\
\text{VII} & \varepsilon, & \varepsilon, & \varepsilon, & \varepsilon, & \varepsilon
\end{array}
$$

We require the number of microstates $t(n)$ given by equation (2.5) for each of these seven distributions.

For distribution $I, n_0 = 4, n_5 = 1$ and all the other ns are zero. So

$$t(n) = \frac{5!}{4!1!0!0!0!0!} = 5.$$

Similarly the other values of $t(n)$ can be evaluated as in the case of $N = 6$ in section 2.3. The seven values of $t(n)$ turn out to be 5, 20, 20, 30, 30, 20 and 1 respectively for the seven distributions. The total of all the $t(n)$s, i.e. the total number of microstates, is thus 126.

EXAMPLE 12.2

Consider a litre of helium gas at $300\,\mathrm{K}$. Use equation (4.3) to calculate n_j for one of its molecules.

The average kinetic energy of one such molecule is

$$\varepsilon_j = \frac{3}{2}k_\mathrm{B}T = 6.21 \times 10^{-21}\,\mathrm{J}$$

using the value of k_B in Appendix 5.
From equation (4.3)

$$\varepsilon_j = n_j^2 \frac{h^2}{8mV^{2/3}}.$$

On the r.h.s. of this equation we have $h = 6.63 \times 10^{-34}$ Js and $m = 6.65 \times 10^{-27}$ kg (see Appendix 5). Also $V = 10^{-3}\,\mathrm{m}^3$. So, on rearranging and substituting in the above equation, we have

$$n_j^2 = 7.5 \times 10^{18}$$

or

$$n_j = 2.7 \times 10^9.$$

We therefore see that the quantum number n_j is extremely large.

EXAMPLE 12.3

A helium atom of mass 6.65×10^{-27} kg is confined to a cubical box of side 1 m.

    (a) Use equation (4.2) to find the energy of the non-degenerate state given by
        $n_x = n_y = n_z = 1$.
        Find also the energy values of the states given by
    (b) $n_x = 3,\ n_y = n_z = 1$,
    (c) $n_x = n_y = n_z = 2$ and

(d) $n_x = 3$, $n_y = 2$, $n_z = 1$,
stating the degeneracy g_j in each case.

From equation (4.2)

$$\varepsilon_j = (n_x^2 + n_y^2 + n_z^2)h^2/8ma^2 = n_j^2 h^2/8ma^2.$$

We have $h = 6.63 \times 10^{-34}$ Js, $a = 1$ m.

(a) For the case $n_x = n_y = n_z = 1$, $\varepsilon_j = \varepsilon_1 = 2.48 \times 10^{-41}$ J on substituting in the above equation.

(b) Substituting now gives us $\varepsilon_j = 11\varepsilon_1$ and since we have three possible combinations, namely

n_x	n_y	n_z
3	1	1
1	3	1
1	1	3

for the same value of ε_j, the degeneracy $g_j = 3$.

(c) We now find $\varepsilon_j = 12\varepsilon_1$ and since we now have the one possibility $n_x = n_y = n_z = 2$, this state is non-degenerate, i.e. $g_j = 1$.

(d) In this case $\varepsilon_j = 14\varepsilon_1$ and we have a degeneracy $g_j = 6$ because of the six possible combinations, namely,

n_x	n_y	n_z
3	2	1
3	1	2
2	3	1
1	3	2
1	2	3
2	1	3

EXAMPLE 12.4

An assembly consists of four particles each of which is in a non-degenerate state of energy 0, ε, 2ε etc. The total energy of the assembly is 6ε.

(a) Write down the possible distributions if the particles are distinguishable. Repeat this for indistinguishable particles which obey (b) the Bose–Einstein statistics and (c) the Fermi–Dirac statistics.

(a) We compile a table, similar to Table 2.2, in which each of n_j particles has an energy $j\varepsilon(j = 0, 1, 2, \dots, 6)$. We see that there are nine possible distributions numbered I, II, ..., IX, thus:

n_0	3	2	2	1	2	1	0	1	0
n_1	0	1	0	2	0	1	3	0	2
n_2	0	0	1	0	0	1	0	3	2
n_3	0	0	0	0	2	1	1	0	0
n_4	0	0	1	1	0	0	0	0	0
n_5	0	1	0	0	0	0	0	0	0
n_6	1	0	0	0	0	0	0	0	0
	I	II	III	IV	V	VI	VII	VIII	IX

Each of the nine vertical columns contains a total of four particles whose total energy is 6ε for each column.

(b) For a BE assembly we also have the above nine distributions because we can have any number of particles in a given energy state (level).

(c) For an FD assembly the only possible distribution is that in column VI because we cannot now have more than one particle in the same energy level; in the other eight columns there are energy levels containing more than one particle.

EXAMPLE 12.5

Evaluate the total number of microstates for case (a) in Example 12.4 and also the average values of the distribution numbers n_0, n_1, etc. for this assembly.

To evaluate the number of microstates for each of the nine distributions in the table of example 12.4 we use the expression on the r.h.s. of equation (2.5) which in this case takes the form

$$\frac{4!}{n_0! n_1! ... n_6!}.$$

Substituting the appropriate values of the ns in each of the nine columns in turn we find that the corresponding numbers of microstates for each of the distributions are

$$\Omega_I = 4, \Omega_{II} = 12, \Omega_{III} = 12, \Omega_{IV} = 12, \Omega_V = 6,$$

$$\Omega_{VI} = 24, \Omega_{VII} = 4, \Omega_{VIII} = 4 \text{ and } \Omega_{IX} = 6.$$

Hence the total number of microstates is $\Omega = 84$.

To find the average value of one of the ns we use a formula like that used to find the centre of gravity of a number of particles, namely

$$\bar{x} = \frac{\sum m_j x_j}{\sum m_j} = \frac{\sum m_j x_j}{M}.$$

In this case the average value of one of the ns is

$$\bar{n}_j = \frac{\sum n_j \Omega_j}{\Omega}$$

Take, for example, the values of n_2 in the third row of the table. The average value is

$$\bar{n}_2 = [1 \times \Omega_{III} + 1 \times \Omega_{VI} + 3 \times \Omega_{VIII} + 2 \times \Omega_{IX}]/84$$

$$= \frac{60}{84} = 0.71.$$

Again, for example, we have

$$\bar{n}_4 = [1 \times \Omega_{III} + 1 \times \Omega_{IV}]/84 = \frac{24}{84} = 0.29.$$

The other $\bar{n}s$ can be similarly found. The complete set of average values is

$$1.33, \ 1, \ 0.71, \ 0.48, \ 0.29, \ 0.14 \text{ and } 0.05.$$

EXAMPLE 12.6

Assuming that the Fermi energy μ is given by equation (7.8) at a temperature $T = T_F/10$ find the value of the Fermi function $f(\varepsilon)$ when $\varepsilon = 0.8\mu_0$.

From equation (7.8), $\mu = \mu_0 \left[1 - \frac{\pi^2}{12} \left(\frac{k_B T}{\mu_0} \right)^2 \right]$. Also $\mu_0 = k_B T_F$ from section 7.2 and $T = T_F/10$. So, on substituting in the above equation,

$$\mu = \mu_0 [1 - \pi^2/1200] = 0.9918\mu_0.$$

The value of $f(\varepsilon)$ is then obtained from equation (7.2), with $\varepsilon = 0.8\mu_0$. Hence

$$f(\varepsilon) = 1/[\exp(-1.918) + 1] = 0.872.$$

EXAMPLE 12.7

An assembly consists of N identical linear oscillators. Find the average fraction n_j/N of oscillators with energy $\varepsilon_j = (j + 1/2)h\nu$ for the three lowest energy levels when $T = \theta$, the characteristic temperature of the assembly.

We use the theory of section 9.2.

From equation (3.3a)

$$\frac{n_j}{N} = \left(\frac{1}{Z}\right) exp(-\varepsilon_j/k_{\mathrm{B}}T)$$

$$= \frac{1}{Z} exp[-(j+1/2)\theta/T]$$

where Z is given by equation (9.1) and $\theta = h\nu/k_{\mathrm{B}}$. Therefore

$$\frac{n_j}{N} = [1 - \exp(-\theta/T)] \exp(-j\theta/T).$$

When $T = \theta$ we get

$$\frac{n_j}{N} = 0.632 \exp(-j).$$

We require the values of n_j/N when $j = 0$, 1, and 2; these values are

$$n_j/N = 0.632, \ 0.232 \text{ and } 0.085.$$

13

Questions, with answers, for the student

The questions in (a) below are mainly numerical examples but there are also descriptive questions, typical of those set in examination papers.

(a) Questions

1. An assembly consists of four distinguishable particles each of which may have an energy $0, \varepsilon, 2\varepsilon, 3\varepsilon$, etc. If the internal energy U is equal to 4ε show that there are 5 distributions and a total of 35 microstates.
2. Find the average value of each distribution number (n_j) in the assembly of Question 1.
3. An assembly is made up of ten localized particles each of which may have an energy $0, \varepsilon, 2\varepsilon, 3\varepsilon$, etc. If the total internal energy is 5ε, discuss how this may be distributed among the particles and show that there are 7 distributions and a total of 2002 microstates.

    What is the number of microstates corresponding to the most probable distribution?
4. Repeat Example 12.1 (Chapter 12) for 5 distinguishable particles but with $U = 4\varepsilon$, showing that there are 5 distributions and a total of 70 microstates.
5. Consider an assembly of localized particles each of which can be in one of three energy states ε, 2ε and 3ε. Write down an expression for the partition function of a particle of the assembly.

What is the ratio of the number of particles in the highest energy state to that in the lowest?

6. Using equation (4.3) find the value of n_j for which an oxygen atom contained in a cubical box of side 1 cm will have the same energy as the lowest possible energy of a helium atom confined to a cubical box of side 2×10^{-8} cm.

7. Write down, in table form, the possible sets of the three quantum numbers n_x, n_y, n_z corresponding to the eight lowest energy levels of a particle in a cubical box of side L. Find the degeneracy g_j of each level and the energy ε_j of each in units of $h^2/8mL^2$.

8. A group of energy levels consists of $g_k = 3$ levels and contains $N_k = 2$ particles. Discuss the possible microstates if the particles obey (a) Bose–Einstein and (b) Fermi–Dirac statistics and hence show that there are 6 microstates in case (a) and 3 in case (b).

9. Using equation (6.2) show that the fraction dN/N of molecules of a Maxwell–Boltzmann gas with speeds between v_m and $1.01v_m$ is 8.3×10^{-3}.

10. A gas consists of eight particles, three of which have speeds of 10 ms^{-1} and five speeds of 20 ms^{-1}. Calculate the average speed $\bar{v}$ and the r.m.s. speed v_r.

11. Find the r.m.s., average and most probable speeds of an oxygen molecule of mass 5.31×10^{-26} kg at 300 K.

12. Using equations (6.17) and (6.18) derive the following result for the vibrational heat capacity of a diatomic gas:

$$C_{Vv} = Nk_{\mathrm{B}}(\theta_v/T)^2 e^{\theta_v/T}/(e^{\theta_v/T} - 1)^2.$$

13. The number of free electrons per cubic metre in copper is 8.5×10^{28}. Using equation (7.7) find the Fermi energy given that $m = 9.11 \times 10^{-31}$ kg for copper. Find also the Fermi temperature.

14. Using the data in Example 13 above find the value of the second term in the bracket in equation (7.8) at a temperature of 300 K. At what temperature does this second term represent a one per cent correction in this equation?

15. The effective mass of electrons in GaAs is less than that of holes. What effect would this have on the position of the Fermi level with respect to the mid-band point of the semi-conductor (a) at room temperature and (b) as the temperature tends to absolute zero?
Explain your answer in (a) by reference to the dependence of $g(\varepsilon)$ on m^* and the dependence of $f(\varepsilon)$ on ε_{F}.

16. (a) A donor impurity is deliberately implanted into a certain semiconductor at a density of 10^{22} atoms m^{-3}. The donor's electron energy states lie 3 mV below the conduction band edge of the semiconductor. The band gap of the semiconductor is 0.8 eV and the effective mass of the conduction band electrons is $0.042\,m_e$, where m_e is the free electron mass. Assuming that the number of holes is comparatively negligible and that $\exp[(\varepsilon_{\mathrm{g}} - \varepsilon_{\mathrm{F}} - 0.003)/k_{\mathrm{B}}T] \gg 1$, find the position of the Fermi level at room temperature.

(*Hint*: The donor atoms contribute 10^{22} electron states per m^3 at $\varepsilon = \varepsilon_g - 0.003$ and the absence of an electron means a charged donor.)

(b) Are the assumptions made above justifiable? (To answer this you need to know that $m_h^* = 0.46 m_e$.)

17. An ideal boson gas consists of zero-spin particles of mass 6.65×10^{-27} kg. The Bose temperature is 0.087 K. Use equation (8.10) to find the number of bosons per cubic metre.

18. Find the fraction of bosons in the ground state of an ideal boson gas at a temperature of $0.1 \, T_B$.

19. Show that the value λ_m of λ for which $u(\lambda)$ in equation (8.16) is a maximum is given by the equation

$$1 - e^{-x} = x/5$$

where $x = hc/\lambda k_B T$.

Hence, by solving this equation, show that $\lambda_m T = 2.9 \times 10^{-3}$ mK.

Calculate λ_m for the earth if one assumes the earth to be a black body.

20. The energy spectrum of sunlight shows a maximum at $\lambda_m = 4840$ Å. If we assume that the sun is a black body, find the temperature of its surface.

21. Repeat Example 12.7 (Chapter 12) for (a) $T = \theta/2$ and (b) $T = 2\theta$.

22. The Debye characteristic temperature for a substance is 1800 K. Find the value of C_V at 200 K using equation (9.9).

23. Using equations (3.12), (3.13a), (9.1) and (9.2) find an expression for the entropy S of an assembly of N identical linear oscillators. Hence show that $S \to 0$ as $T \to 0$.

24. If the mutual potential energy of two molecules is given by

$$\phi(r) = \frac{B}{r^{12}} - \frac{A}{r^6}$$

(a result due to Lennard–Jones and Devonshire in 1937), find the values of ε, σ and r_0 in Fig. 10.2 in terms of A and B.

25. If, in the previous example, the distance between the two molecules is increased from the normal equilibrium separation, r_0, to $2r_0$, show that their mutual potential energy then becomes $\phi(r) = -\varepsilon/32$.

26. If the mutual potential energy of two molecules is given by the Mie potential

$$\phi(r) = \frac{-A}{r^m} + \frac{B}{r^n}$$

(a result due to Mie in 1907), express B in terms of A, m, n and r_0 (see Fig. 10.2). Hence rewrite this potential in a form which does not contain B.

27. Discuss the statistics of an assembly of non-localized particles explaining what is meant by symmetric and antisymmetric wave functions and indicating their relationship with the Bose–Einstein and Fermi–Dirac statistics.

28. Explain why it is necessary to group the various available energy states into 'bundles' when dealing with the statistics of an assembly of non-localized particles.

    How does this description of such an assembly differ from that of an assembly of localized particles?

29. Explain what is meant by a partition function and derive it for the translational motion of a molecule of a perfect gas.

    Show how this result can be extended to construct a partition function for a simple liquid.

30. 'The object of statistical mechanics is to derive the properties of matter in bulk from the known or assumed structure of matter and the laws of interaction between its constituent particles.'

    Discuss this statement.

31. Distinguish between (a) an assembly of localized particles and (b) one of non-localized particles, giving examples of each kind.

    Derive the distribution function for an assembly of type (a).

32. Write concise explanatory notes on the following topics:

    (a) the relation between entropy and probability;

    (b) the partition function;

    (c) the total number of microstates for an assembly of localized particles.

(b) Answers to questions

2. 1.71, 1.14, 0.69, 0.34 and 0.11.

3. 840.

5. $Z = e^{\beta\varepsilon} + e^{2\beta\varepsilon} + e^{3\beta\varepsilon}$; ratio is $e^{2\beta\varepsilon}$.

6. $\sqrt{3} \times 10^8$.

7.

n_x	n_y	n_z	g_j	ε_j
1	1	1	1	3
2	1	1	3	6
2	2	1	3	9
3	1	1	3	11
2	2	2	1	12
3	2	1	6	14
4	1	1	3	18
3	3	1	3	19

10. $\bar{v} = 16.25$ ms^{-1}, $v_r = 16.95$ ms^{-1}.

11. $v_r = 482$ ms^{-1}, $\bar{v} = 445$ ms^{-1} and $v_m = 394$ ms^{-1}.

13. 1.13×10^{-18} J, 8.2×10^4 K.

14. 1.1×10^{-5}, $T = 9040$ K.

16. (a) 0.72 eV.

17. 10^{26} m^{-3}.

18. 0.97.

19. 1.06×10^{-5} m.

20. 5990 K.
21. (a) 0.865, 0.117, 0.016.
    (b) 0.394, 0.239, 0.145.
22. 2.65 J mole^{-1} K^{-1}.
24. $\varepsilon = A^2/4B$, $\sigma = (B/A)^{1/6}$, $r_0 = 1.12(B/A)^{1/6}$.

Appendix 1:
Stirling's approximation

Since $N! = 1 \times 2 \times 3 \ldots \times N$ then

$$\ln N! = \ln 1 + \ln 2 + \ln 3 + \ldots + \ln N$$

$$= \sum_{1}^{N} \ln x$$

= the sum of the areas of the rectangular strips

of unit width up to $x = N$ in the histogram

of Fig. A1.1

$\simeq$ for large N, the area under the curve of

$\ln x$ between $x = 1$ and $x = N$.

So

$$\ln N! \simeq \int_1^N \ln x \, dx$$

$$= [x \ln x - x]_1^N$$

$$= N \ln N - N + 1$$

$$\simeq N \ln N - N \qquad (A1.1)$$

when $N \gg 1$.

Stirling gave the more accurate approximation

$$\ln N! = N \ln N - N + \frac{1}{2} \ln 2\pi N$$

but the result (A1.1) is perfectly satisfactory for the large values of N with which we are concerned in statistical mechanics.

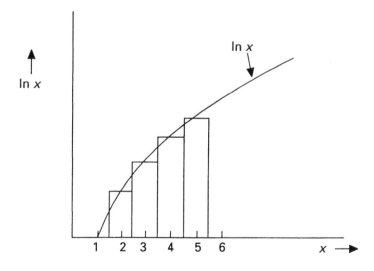

Fig. A1.1. Histogram of $\ln x$ for integral values of x together with the continuous curve of $\ln x$.

Appendix 2:
The result $\Omega = N!/\prod_j n_j!$ for
localized particles

Consider first the number of ways in which we can place N localized (distinguishable) particles into three boxes labelled $1, 2$ and 3 such that there are n_1, n_2 and n_3 particles respectively in these three boxes and where $N = n_1 + n_2 + n_3$.

The number of ways of choosing n_1 particles out of N for box 1 is

$$NC_{n_1} = \frac{N!}{n_1!(N - n_1)!}. \tag{A2.1}$$

We are then left with $(N - n_1)$ particles and we can therefore choose n_2 of them for box 2 in

$$(N - n_1)C_{n_2} = \frac{(N - n_1)!}{n_2!(N - n_1 - n_2)!} \tag{A2.2}$$

ways.

Finally the number of ways of putting the remaining n_3 particles into box 3 is clearly just one.

So the number of ways Ω of choosing particles to occupy the three boxes is the product of the right-hand sides of (A2.1) and (A2.2), that is,

$$\Omega = \frac{N!}{n_1! n_2! (N - n_1 - n_2)!}$$

(A2.3)

$$= \frac{N!}{n_1! n_2! n_3!}.$$

So for the more general case $N = n_1 + n_2 + \ldots + n_j + \ldots$ the number of ways in which we can place N distinguishable particles into boxes labelled $1, 2, \ldots, j, \ldots$ so that there are $n_1, n_2, \ldots, n_j, \ldots$ particles in the boxes is an extension of (A2.3) and we have

$$\Omega = \frac{N!}{n_1! n_2! \ldots n_j! \ldots}$$

(A2.4)

$$= \frac{N!}{\prod_j n_j!}.$$

This is the result we met in equation (2.5).

Appendix 3: Various integrals

In this appendix we gather together the various integrals that have appeared in the text.

The integrals

$$\int_0^\infty x^3 e^{-ax^2} dx = \frac{1}{2a^2}$$

and

$$\int_0^\infty x^4 e^{-ax^2} dx = \frac{3}{8a^2} \left(\frac{\pi}{a}\right)^{1/2}$$

were encountered in section 6.2 when discussing the distribution of molecular speeds.

We used the result

$$\int_0^\infty \varepsilon^{1/2} e^{-\lambda\varepsilon} d\varepsilon = \pi^{1/2}/2\lambda^{3/2}$$

when calculating the partition function for the translational motion of one particle of a Maxwell–Boltzmann gas (section 6.3).

The other three integrals which we met are

$$\int_0^\infty \frac{x^{1/2}}{e^x - 1} dx = \frac{2.612\pi^{1/2}}{2} = 2.315 \qquad \text{(see section 8.2)}$$

$$\int_0^\infty \frac{x^3 dx}{e^x - 1} = \frac{\pi^4}{15} \qquad \text{(see section 8.5)}$$

and $$\int_0^\infty \frac{x^4 e^x}{(e^x - 1)^2} dx = \frac{4\pi^4}{15} \qquad \text{(see section 9.4)}.$$

Appendix 4:
Some thermodynamic results

Consider an assembly of N particles. We assume that the state of the assembly can be described by P, V and T (the state variables). The assembly can be treated from the point of view of thermodynamics or statistical mechanics and there are various equations which form a 'bridge' between the two approaches. Perhaps the most celebrated is

$$S = k_B \ln \Omega. \tag{A4.1}$$

Equations (A4.14) and (A4.15) are further examples of such equations.

From now on we concentrate on the thermodynamic approach. The assembly has certain *extensive properties* which are proportional to the size of the assembly, that is, to N. Examples of extensive properties are the volume V, entropy S and internal energy U. Other properties, known as *intensive properties*, are independent of the size of the assembly, examples being the external pressure P and temperature T.

Suppose that, in a small change of state, a quantity of heat dQ is absorbed by the assembly. Then by the first law of thermodynamics we can write

$$dU = dQ + dW. \tag{A4.2}$$

Here dU is the change in internal energy and $dW = -PdV$ is the work done on the assembly. Further, if the change is reversible, then by the second law of thermodynamics

$$dQ = TdS \tag{A4.3}$$

and so from (A4.2) and (A4.3) we have

$$dU = TdS - PdV. \tag{A4.4}$$

This is often known as the *central equation* of thermodynamics.

We now define three other quantities—the enthalpy H, the Helmholtz function (or Helmholtz free energy) F and the Gibbs function G. They are

$$H = U + PV \tag{A4.5}$$
$$F = U - TS \tag{A4.6}$$
$$G = F + PV = U - TS + PV. \tag{A4.7}$$

(We have not explicitly used H and G in the text.)

We now consider the differences in the values of these functions in two neighbouring equilibrium states of a *closed PVT* system or assembly. (A closed system is one in which the amount of matter is constant, that is, no matter enters or leaves the system—see section 5.5.)

We have

$$dU = TdS - PdV \tag{A4.4}$$
$$dH = dU + PdV + VdP$$
$$dF = dU - TdS - SdT$$
$$dG = dU - TdS - SdT + PdV + VdP.$$

Using (A4.4) we eliminate dU in the last three equations to give

$$dH = TdS + VdP \tag{A4.8}$$
$$dF = -SdT - PdV \tag{A4.9}$$
$$dG = -SdT + VdP. \tag{A4.10}$$

Consider first equation (A4.4). If we regard U as a function of S and V then

$$dU = \left(\frac{\partial U}{\partial S}\right)_V dS + \left(\frac{\partial U}{\partial V}\right)_S dV$$

and comparing this with (A4.4) gives us

$$T = \left(\frac{\partial U}{\partial S}\right)_V, \quad P = -\left(\frac{\partial U}{\partial V}\right)_S. \tag{A4.11}$$

We can use the same procedure for dH, dF and dG. For example

$$S = -\left(\frac{\partial F}{\partial T}\right)_V, \quad P = -\left(\frac{\partial F}{\partial V}\right)_T. \tag{A4.12}$$

We next consider an *open* system in which the quantity of matter (i.e. the number of particles N) is not fixed. So instead of equation (A4.4) we must write

$$\mathrm{d}U = T\mathrm{d}S - P\mathrm{d}V + \mu\mathrm{d}N \tag{A4.13}$$

to allow for the fact that N is no longer constant. The quantity μ is the chemical potential which we met in section 5.5.

Suppose now that we picture our final assembly as being built up in stages from nothing in such a way that the intensive properties T, P and μ on the r.h.s. of (A4.13) remain constant. Then we need to sum the various equations of type (A4.13) over all these stages to obtain our equation representing the final state of the assembly, so we write $\Sigma \mathrm{d}U = U$, $\Sigma \mathrm{d}S = S$, $\Sigma \mathrm{d}V = V$ and $\Sigma \mathrm{d}N = N$. Thus we obtain

$$U = TS - PV + \mu N$$

or

$$U - TS + PV = G = \mu N.$$

This shows that μ is the Gibbs free energy per particle, as long as only *one* type of particle is present.

At the beginning of this appendix we cited equation (A4.1) as an example of a 'bridge' equation. Two further such examples are

$$F = -k_{\mathrm{B}}T \ln Z \tag{A4.14}$$

and

$$U = k_{\mathrm{B}}T^2(\partial \ln Z/\partial T). \tag{A4.15}$$

On the l.h.s. of each of the three equations we have a thermodynamic quantity and on the r.h.s. we have a quantity in statistical mechanics.

For further reading see *Thermal physics* by C. B. P. Finn (Routledge and Kegan Paul, 1989).

Appendix 5: Physical constants

Quantity	Symbol	Value
Planck constant	h	6.63×10^{-34} Js
Boltzmann constant	k_B	1.38×10^{-23} JK^{-1}
Avogadro constant	N_A	6.02×10^{23} mol^{-1}
Gas constant	R	8.31 JK^{-1} mol^{-1}
Electronic charge	e	1.60×10^{-19} C
Mass of electron	m_e	9.11×10^{-31} kg
Mass of proton		1.67×10^{-27} kg
Bohr magneton	μ_B	9.27×10^{-24} JT^{-1}
Permeability of free space	μ_0	$4\pi \times 10^{-7}$ Hm^{-1}
Stefan constant	σ	5.67×10^{-8} Wm^{-2} K^{-4}
Speed of light	c	3.00×10^{8} ms^{-1}
Molar volume of an ideal gas at STP		22.4 l
Atmospheric pressure		1.01×10^{5} Nm^{-2} (or 1.01×10^{5} Pa)
Mass of $^4_2$He atom		6.65×10^{-27} kg
Mass of $^3_2$He atom		5.11×10^{-27} kg
1 electron volt (1 eV)		1.6×10^{-19} J

References

CHAPTER 2

Books

Finn, C. B. P, 1989, *Thermal physics*, Routledge, London.
Rushbrooke, G. S., 1949, *Statistical mechanics*, Oxford University Press. (Chapter II).
Schroedinger, E., 1948, *Statistical thermodynamics*, Cambridge University Press.
Wilks, J., 1961, *The third law of thermodynamics*, Oxford University Press.

Papers

Offenbacher, E. L., 1965, *Amer. J. Phys.*, **33**, 950.
Trevena, D. H., 1964, *Amer. J. Phys.*, **32**, 790.

CHAPTER 3

MacDonald, D. K. C., 1963, *Introductory statistical mechanics for physicists*, John Wiley, New York (p. 25).

CHAPTER 4

Tabor, D., 1979, *Gases, liquids and solids*, Cambridge University Press.
Wilks, J., 1961, *The Third law of thermodynamics*, Oxford University Press (Chapter II).

CHAPTER 5

Guenault, T., 1988, *Statistical physics*, Routledge, London (Chapter 5).

MacDonald, D. K. C., 1963, *Introductory statistical mechanics for physicists*, John Wiley, New York (pp. 78 and 152).

Matthews, P. T., 1963, *Introduction to quantum mechanics*, McGraw-Hill Publishing Company, New York (Chapter 8).

CHAPTER 6

Part I

Finn (see Chapter 2 reference list).

Nash, L. K., 1968, *Elements of statistical thermodynamics*, Addison-Wesley, London (Part II).

Pendlebury, J. M., 1985, *Kinetic theory*, Adam Hilger, Bristol.

Rosser, W. G. V., 1986, *An introduction to statistical physics*, Ellis Horwood, Chichester (Chapter 5).

Rushbrooke, G. S., 1949, *Statistical mechanics*, Cambridge University Press (Chapter III).

Tabor, D., 1979, *Gases, liquids and solids*, Cambridge University Press.

Part II

Gopal, E. S. R., 1974, *Statistical mechanics and properties of matter*, Ellis Horwood, Chichester (Chapter 2).

Rosser (see Part I above) (Chapter 5).

CHAPTER 7

Guenault (see Chapter 5 reference list).

Kittel, C., 1976, *Introduction to solid state physics*, John Wiley, New York.

MacDonald, D. K. C., 1963, *Introductory statistical mechanics for physicists*, John Wiley, New York (p. 85).

Rosenberg, H. M., 1989, *The solid state*, Oxford University Press.

Rosser, W. G. V., 1982, *An introduction to statistical physics*, Ellis Horwood, Chichester (Chapter 12).

Wilks, J., 1961, *The third law of thermodynamics*, Oxford University Press (Chapters IV and VII).

CHAPTER 8

Kittel, C., 1958, *Elementary statistical physics*, John Wiley, New York (p. 96).

MacDonald (see Chapter 7 reference list) (Chapter 3)

Rosser, W. G. V., 1986, *An introduction to statistical physics*, Ellis Horwood, Chichester (Chapter 12).

Paper

London, F. 1938, *Phys. Rev*, **54**, 947.

CHAPTER 9

Guenault, T., 1988, *Statistical physics*, Routledge, London (Chapter 3).
Kittel, C., 1976, *Introduction to solid state physics*, John Wiley, New York.

CHAPTER 10

Books

Mayer, J. E. and M. G., 1977, *Statistical mechanics*, Second edition, John Wiley, New York.
Pryde, J. A., 1966, *The liquid state*, Hutchinson University Library, London.
Rushbrooke, G. S., 1949, *Statistical mechanics*, Oxford University Press (Chapter XVI).
Temperley, H. N. V., 1956, *Changes of state*, Cleaver-Hume Press, London.
Temperley, H. N. V. and Trevena, D. H., 1978, *Liquids and their properties*, Ellis Horwood, Chichester.
Temperley, H. N. V., Rowlinson, J. S. and Rushbrooke, G. S., (Eds), 1968, *Physics of simple liquids*, North-Holland, Amsterdam.
Trevena, D. H., 1975, *The liquid phase*, Wykeham Science Series, London.
Trevena, D. H., 1987, *Cavitation and tension in liquids*, Adam Hilger, Bristol.

Papers

Alder, B. J. and Wainwright, T. W., 1957, *J. Chem. Phys.*, **27**, 1208 and also 1959, *J. Chem. Phys.*, **31**, 459.
Kirkwood, J. G., 1935, *J. Chem. Phys.*, **3**, 300.
Percus, J. K. and Yevick, G. J., 1958, *Phys. Rev.*, **110**, 1.

CHAPTER 11

Dugdale, J. S., 1966, *Entropy and low temperature physics*, Hutchinson University Library, London (Chapter 8).
Gopal, E. S. R., 1974, *Statistical mechanics and properties of matter*, Ellis Horwood, Chichester (Chapter 2).
Wilks, J., 1961, *The third law of thermodynamics*, Oxford University Press (Chapter III).

Index

Printed and bound by CPI Group (UK) Ltd, Croydon, CR0 4YY

03/10/2024

01040437-0017